BUG

VOICES FROM THE UNDERSIDE OF SILICON VALLEY

EDITED BY RAJ JAYADEV AND JEAN MELESAINE

HEYDAY, BERKELEY, CALIFORNIA

Library of Congress Cataloging-in-Publication Data

De-bug : voices from the underside of Silicon Valley / edited by Raj Jayadev and Jean Melesaine.
    pages cm
  ISBN 978-1-59714-319-6 (pbk. : alk. paper) -- ISBN 978-1-59714-339-4 (e-pub) -- ISBN 978-1-59714-340-0 (amazon kindle)
  1. Service industries workers--California--Santa Clara Valley (Santa Clara County) 2. Service industries--California--Santa Clara Valley (Santa Clara County) I. Jayadev, Raj, editor. II. Melesaine, Jean, editor.
  HD8039.S45D43 2016
  331.7'930922794'73--dc23
                                     2015019685

Cover Art: Samuel Rodriguez
Cover Design: Ashley Ingram
Interior Design/Typesetting: Rebecca LeGates

Orders, inquiries, and correspondence should be addressed to:
        Heyday
        P.O. Box 9145, Berkeley, CA 94709
        (510) 549-3564, Fax (510) 549-1889
        www.heydaybooks.com

Printed by McNaughton and Gunn, Saline, MI

10 9 8 7 6 5 4 3 2 1

This book is dedicated to:

Donae who sings his song
Demetrius who spits his rhyme
Chilo who watches over our home
Albert who guides us to justice
Kevin who calls on us to always "write about it"

CONTENTS

PREFACE

When we started writing this book, the generation of young adults that truly made Silicon Valley what it is now—through their labor, culture, and struggles—was invisible. You haven't really seen the fuller reality of Silicon Valley until you've seen it through the eyes of a teenage assembler, an undocumented worker building mansions, or even a guy hustling drugs to techies while living under bridges. The true Silicon Valley is more complex than the region's branding and marketing minds will ever disclose. Rather than a collection of pieces written by journalists or academics observing Silicon Valley from the outside, this book brings together the raw experiences of people living and working here. As such, it is as much an unauthorized diary of Silicon Valley as it is a history book.

Most of these writings originally appeared in our magazine, *De-Bug,* which was born when many of the people who wrote for this book were working entry-level temp jobs in assembly plants and manufacturing warehouses. When Sandy Close, director of New America Media (and longtime mentor of ours) asked us to write for their Bay Area youth magazine *YO! (Youth Outlook),* we put our stories down on paper, even though none of us had any journalistic experience. We didn't have an office, so we would meet up at a Vietnamese restaurant in downtown San Jose, which would get awkward because we rarely ordered anything besides water. Those conversations evolved into our first stories.

YO! printed our writings on a four-page spread that read "Young and Temporary in Silicon Valley" and had a drawing of a computer with tentacles piercing the body of a young man. The image illustrated the story by Edward Nieto, one of our founding writers, called "Blood-Sucking Machines," about his job at a medical equipment production company that paid workers to donate blood to calibrate the new products. At the time, Ed was the only one of us who worked at a medical equipment plant, but the story of literally giving your blood to your employer resonated with most of us.

Silicon Valley at the turn of the millennium was a mythological place. It existed in the public imagination as a place of never-ending opportunity—an overflowing lottery of prosperity—and the magnetism of its image drew families from all over the world. Silicon Valley's blueprint was so appealing that regions all across the country began to rename themselves to be part of the Silicon fraternity—Silicon Alley, Silicon Forest, Silicon Hills.

From the inside of the machinery, Silicon Valley was a starkly different place than the futuristic utopia it was projected to be. When the image boasted high incomes, most people just saw high rents. When it said more million-dollar homes, we saw more people becoming homeless. Young people were living in parallel universes hidden behind the one seen on the evening news. We wanted to create a space for people to articulate this world in their terms.

So we kept meeting up and growing our circle of storytellers. We wanted to find a name for our group that people would know from their own lives. *De-bug* was a word many of our authors heard every day at work in the back end of Silicon Valley's big-name companies—the HPs, Ciscos,

Intels—stuffing printers into boxes, inserting formatters, cutting microchips, and doing all the other crucial but invisible jobs that gave birth to the high-tech age. If a product wasn't working right at the end of the line, you called in the "De-Bug" unit to inspect the product and expose the malfunction in order to correct it. Liz Gonzalez, one of our founding writers, offered up the name when we were figuring out what to call our fledgling little operation. It resonated because it came from that language of personal experience, and it was appropriate since we were trying to expose the malfunctions of the Silicon Valley economy and politics.

We became the distributors of *YO!* in Silicon Valley. Victor Saldana, another founding writer, had the idea of following lunch trucks when they drove into the back parking lots of the tech companies and handing out the magazines when people were changing shifts or on lunch breaks. If we saw smocks and hairnets instead of ties and suits, we knew we were in the right place. People would pick up a copy of our magazine along with their fried burritos. We got chased out of many parking lots by security, but people were feeling it, mainly because it was the first time they were seeing their own lives reflected in print.

Usually, when we handed someone a *De-Bug* magazine, they had an interesting story of their own but didn't feel they were qualified to write it since they weren't "a journalist." We'd tell them they were the best person to write the story because they were the one who had experienced it, and that no one could offer the perspective they could. We had more people who wanted to write than we had room for in *Youth Outlook,* so we decided to create our own magazine. New America Media helped us to get off the ground and has our back to this day, publishing our pieces on their national newswire.

The reflective experience of writing also led to other ambitions: not only to name the world we live in, but also to change it. We encouraged any and all individual aspirations. Our question evolved from "What would you like to write about?" to "What would you like to do?" In that way, we learned to operate more like a community than an organization. Even to this day, we call ourselves an "un-organization" because rather than people leaving their personal goals aside for the good of the group—as people would in most conventional teams, businesses, or organizations—we do the opposite. People come into the De-Bug circle and lead us in whatever directions feel right to them.

That sort of freedom, to determine your own future through your own audacious aspirations, is actually what Silicon Valley is all about. We just said that the invitation should be given to all, that the "Silicon Valley dream" should be democratized. Today's Valley speak would call it "open-sourcing." And yet despite the varied directions in people's lives and aspirations, we have stayed close as a family. We still have an open-door story-exchange meeting once a week that anyone can attend, just like we did in that restaurant years ago.

Ultimately, Silicon Valley is not limited to geography, or even industry. To most of the world, it is a promise more than a place, a dream in which innovation and creativity reign. Ironically, you will find this to be as true, if not truer, in the stories of people working odd or anonymous jobs as you do in the stories of the famous CEOs and venture capitalists. This book is a testimony to the power of everyday people telling their own stories. We all have the capacity to reflect on our lives and communicate the complexity, brilliance, and poetry of our experiences to one another. That investigation, as it turns out, also produces the most profound, hilarious, heart-breaking, shocking, and illuminating stories imaginable.

Communities—societies, actually—can form in the shadows of a region's public image. In the process of writing this book, and the years of work that led to it, this is perhaps the most powerful byproduct we found of people telling their stories to one another: the most unlikely people can find common ground to build relationships. We hope reading this book transforms you in some way, because the journey we took in creating it has certainly transformed us.

ANGEL LUNA

EL PALETERO

It is seven in the morning, and I start my day by cleaning
my workstation—my ice cream cart. I am part service pro-
vider, part entrepreneur, and what most people here in East
San Jose call a paletero—the neighborhood ice cream man.

   I have my supplies ready, and I check the weather for
one of the most important ways to increase sales: a hot day.
In some ways, the job is straightforward—push, walk, and
more walking and more pushing. Having an entrepreneur-
ial spirit is one of the things that made me get in the business of cold sweet treats. Roaming the empty streets during
morning hours is part of the hustle. Sometimes it's so dead
that I'm the only one cruising these empty streets, just
wringing my bells loud enough to let people know I'm in
their streets, ready for business. The bells are a part of the
soundtrack of any working-class neighborhood in California. After the sound waves reach their ears, a few people
come out to meet me for a cold treat. As I greet them with
a warm smile and a hello, they usually have already made
up their minds about what they want. But some are just
plain indecisive and can't make up their minds about the ice
cream they want. For them, it's part of the experience—the
searching through the cart for their elusive treasure.

I consider being a paletero the tastiest "green job" in Silicon Valley. And while it can be fun and fulfilling (you are a hero to children), not everyone is cut out for this work. The walking is demanding, and the amount of miles that I put on my shoes is a reflection of my hustle. The troubles out there at your workplace, the streets, are also very real. You have to be careful due to the fact that someone might mistake you for a walking ATM. It can get very dangerous in certain areas of the city, to the point that you might have to be ready to defend yourself or to make sure they don't take your earnings of the day. The risk lives at any corner and pretty much in anything that might be hidden out there in the streets. But for the most part, the gangsters respect your hustle, because they might have seen you long enough in the streets that they know you're not stepping on toes or there to take nothing from them. The hood is always going to be the hood, so never take it personal. Life happens and you just have to adjust or learn how to deal with the situation that it may present in the daily basis.

The cops sometimes can be a hassle and very disrespectful because of the street peddler status of paleteros. Not all of them are bad, but a lot of them look down on you because you're not in a fancy cubicle or doing what we think is a good job. Their usual gripe is, "You can't be selling here." Some cops are just looking for any excuse to search you or to look for a probable cause to put you down. They know that the majority of paleteros are immigrants that might not have the proper documentation, so to them we are easy targets to harass.

The challenges, though, are not only cops and robbers—it's also the clock. The job is a race against time because once the dry ice and the ice cream are in the cold box, the stopwatch starts. So that means that break times, lunch, anytime you are not selling is costing you money. To add to that struggle, people don't always have enough to pay at the sticker price. You just got to work with them; this game is a give and take. Sometimes your product might be a bit damaged or it's that point of the day that you have to liquidate the inventory—mainly because it is already turning into liquid.

And the competition can also be pretty fierce at times, though I never looked at it that way. When you see another fellow paleta man, the first thing you do is give a heads-up on what's going on in your route—any danger, hot spots like a birthday party at the park, or people to avoid. Some can be greedy with a spot, and they try to outdo you. There have been a number of times when I've seen some paleteros, and rather than say hi, they sprint ahead to jump my route. But I just keep it pushing, and let them know that there's enough pie (or ice cream) for everyone.

By the end of the day, my legs are exhausted, and my body is overheated from the fierce sun. I store any remaining product for the next day. The earnings might not be the equivalent to the hard work that you put into the trade, but the satisfaction of making someone's day from a walking freezer has no price.

DANIEL ZAPIEN

REFLECTIONS OF A POOL BUILDER

Growing up in California, I always wondered how people could afford to have underground pools in their homes. Today, I work for a pool construction company, and as I demolish some pools and dig holes for new pools, I am getting a window into who's swimming and who's drowning in the Silicon Valley economy.

Having done this for over a year, the pattern I see is that those who want pools made are new money—young techies on their way up. The ones demolishing are old money—former bosses of companies and industries, some that don't exist anymore. The truth is that regardless of the unemployment rate, or the stock market, in Silicon Valley there are some that are coming up and some whose times have passed. I see who they are while digging holes in their backyards. Based on the type of pools I have to dig, and the locations we go to, I can see why I never had one growing up. This business is expensive. The price for renting the tractors, the wood, the steel, the cementer—it all gets pricey, and I'm only the first step. There are still the landscapers, the steel guys, the permits, and more. The average pool will run anywhere from $25,000 to $50,000.

I go all over Silicon Valley seeing these pools built and demolished. The craziest one I saw was in San Jose. I didn't even know there were huge homes like that on my side of the city. New money in an old part of town.

Most clients who want their pool demolished, first off, have a dog or two, are over fifty, and have kids who have moved out of the house. Those who want a pool built, on the other hand, are up and coming. They either already have money or started making lots of it through one of the many jobs now flourishing in Silicon Valley. The differences between the two groups—in age, outlook, prospects—are pretty clear.

I met one woman after demolishing a pool at her house in Santa Clara. When we finished she cooked up some BBQ and began telling us stories of growing up on a farm. She talked about how living the city life was different than what she was used to. She said she missed talking to people in person, like she did when she was a kid, versus all the online communication that happens now.

I met another guy, an architect, who told me about how he came to his status in life. I was still new on the job, and he came up to show me how to water the mounds of dirt in his yard to keep it from getting dusty. He told me his secret was that he was a leader and not a boss. He helped his workers when times were hard, like digging with them and staying long after shifts were over to make sure the job was done correctly.

My own boss, Jake, treats us pretty fairly. "I won't make you do anything unless I either did it or do it now," he tells us. But when the economy sank and work dried up, even that wasn't enough to keep my coworker, Alvin, from leaving the company.

Alvin is in his forties and from Mexico. He's one of the hardest working people I know, one of those guys you can joke around with at the workplace but still get the job done. He always sported his hat backwards and was the main driver for the bulldozer. He had a million stories, and told me about his working days before working with our boss, Jake.

Alvin rents an apartment in the East Side of San Jose and lives with some of his cousins. While digging pools, he also worked another job at the time, putting stucco siding on houses. All of that was to support his daughter, who recently had a baby, his nephew, who is going to college full time to be an accountant, and his cousins, all of who work as hard as he does. But like a lot of immigrant families, Alvin lives check to check, so when the pay began to dip he left to find other work.

I've managed to stay with the company, and business is starting to pick back up. And yes, one day, I do want to own my own pool. I want the life of the people who swim in Silicon Valley.

ANTHONY KING

ADDICTION AND REDEMPTION IN SILICON VALLEY

At the height of Silicon Valley's economy, I was a quality assurance engineer at a molding company in Santa Clara. I was responsible for the inspection of manufactured parts for an international assembly line. One summer, my employer relocated to the Oregon border, and while my pay was substantial, it wasn't enough for me to negotiate the challenges of relocating.

I found myself bouncing from temp job to temp job, from inspecting circuit boards to game testing for Electronic Arts. I was trying to secure a position in a field that was slowly dwindling in this area. While I struggled finding a job, I had a lot of free time on my hands and began hanging out with friends who made fast money selling drugs.

I began to hear the sirens' call of easy money. In my naiveté I used three thousand dollars to "invest" in the business of selling crack cocaine. I promised myself this was just a "side job," something to augment the meager earnings

I was making temping, and vowed that as soon as I found full-time employment, I would immediately leave this life of crime. Things looked very bright. I had a business model, a two-year exit plan for my new job in drug sales, and I was equipped with the same meticulous approach I used when I inspected parts on the assembly line. I was able to maintain my quality of life, even elevate it in some aspects. I had enough money to pay my mother's rent, maintain my own apartment, and lead an active social life.

The majority of people view narcotics users as folks down on their luck, with broken family structures or some type of social shortcomings. However, that wasn't true. My clients were from all parts of the economic ladder. Some came from blue-collar backgrounds, like machine operators from local turnkey manufacturers, construction workers, and even a county bus driver. I also had quite a few white-collar and tech workers: assembly workers for tech manufacturers, and managers and engineers from such places as Hewlett Packard, and even a middle-school teacher. I was even able to secure a temp job at a tech firm from one of my clients. But as that first year came to a close, I was beginning to lose control. I'd developed an addiction to the same garbage I sold, and it creeped in slowly. I began smoking "chewys," marijuana blunts laced with crack. In a matter of months my casual use rapidly blossomed into a full-fledged habit. It was beginning to affect my work, both in the office and on the streets. In the office, absenteeism began to affect my productiveness. On the street, my "slip was showing," so to speak, and it began to cost me credibility and customers.

My quality of life spiraled, and addiction convinced me nothing was wrong. This disconnect from reality would cause me to lose all my earthly possessions, including my apartment. I was literally opening the door to life on the streets.

My homeless tenure began simple enough. A small segment of my customer base lived along the Guadalupe River, which weaved through San Jose. They took me in. It was awkward at first, but slowly I adjusted. It took a lot of getting used to, going from a one-bedroom apartment to a four-by-six-foot tent. And I saw things I had never seen before.

One of my first nights there I was awakened by my camp mates having a mild but tense discussion. I peeked out of my tent and I saw my camp mate "Dog" seated with another camper, John, holding him in a headlock with a vice-like grip. John's wife, Rita, was standing over them with a hypodermic syringe aimed at Dog's neck. I was just about to shout "What's goin' on?" A woman I shared my tent with covered my mouth and cautioned me to be quiet. "You're gonna break their concentration," she said. I watched in awe and dismay as Rita emptied the syringe into Dog's neck. I'd later learn that Dog was a heroin addict. His veins had collapsed in his arms and legs through years of his narcotics use. His neck was the only sweet spot for him to get his fix.

Living on the street, the tech sector's long arm reached my way. I had another neighbor "SD" who had a number of clients from Adobe who would come to our encampment looking for crack. I would marvel at how these software gurus (some of whom would wear their Cole Haan shoes and J. Blades neckties to our humble freeway encampment) would lay on mats, like preschoolers at daycare. Titans of tech would look so impotent as SD spoon-fed them mere crumbs of the narcotic, knowing full well that they had overpaid.

One day some Fish and Game wardens were at the river where we lived, handing out citations and assertively "suggesting" that residents relocate. I wasn't there, so instead of giving me a ticket, they made an example of my tent. A game warden attached a rope to my tent, dragging my tent about thirty feet. When I returned the next day, I found not just my tent gone, but all of my neighbors along that stretch of the river gone too. I found them in an empty field about half a mile from our old spot.

Everyone knew what had happened to my tent. My friends gave me another one, along with some of my belongings they had managed to salvage. It would take me a month to catch up with the game warden that had destroyed my tent, and he was very unyielding in his opinion that he have charges filed against me for illegally camping. The D.A. later dropped the charges against me. I saw this as a sign of things to come.

The spot of our new encampment was very open, covering nearly a whole city block. We were visible to everyone, from the various groups in the faith-based community that came to offer food and spiritual support, to the nonprofit outreach organizations that came offering assistance for getting off the streets. While the outreach workers offered a path that could lead to stability, the church groups gave me something much more: a reason to have faith!

In March I applied for a program that was designed to provide chronically homeless individuals subsidized housing. A few days later the city would execute a sweep of that site. Driving the 250-plus residents in all directions, except toward the river, that is.

In May I was accepted into the aforementioned program and began to work with a case manager. On August 1, the Housing Authority issued me a Housing Choice Voucher (commonly known as Section 8). I had been informed the previous week that I would be getting my voucher, and decided to celebrate by getting high (as if I needed an excuse).

As I began to prepare my "issue," I pondered all the things I would be able to do after I secured a place to live. I was just about to slide the needle into my vein when the thought hit me: "You keep putting a needle in your arm, and then the only thing you will be is a junkie with an apartment!" That very day I stopped using needles. Some would say "That's no big deal," but to me it was. You see, I felt 90 percent of kicking my addiction would depend on me leaving needles alone.

In my weekly meetings with my grandmother, I would tell her about these issues. She would give sage reassurance and tell me to pray.

I had not prayed, earnestly, in over seven years. Later that week I received notifications from three separate complexes refusing to rent to me. I returned to my tent dejected and bitter. I got high.

I began to sob uncontrollably and continued to cry most of the night. I noticed a stack of mail my grandmother had given me and began to recall her advice. I opened a travel version of the book of Psalms. The only reason I had this bible was to use the pages as rolling paper. As I thumbed through the half-torn pages I came across a swatch of paper with a bible verse written in crayon. Psalm 100. After reading it a few times for clarity, I closed my eyes and asked the Lord to save his lost sheep. I promised the Lord that if he saved me, I would do my best to be worthy of his grace. The following week I would meet a woman that worked in the field who suggested I transfer to another case management group. With her tutelage and mediation by the city's housing department, and the grace of God, my case manager and I were able to secure my own apartment.

To say my quality of life has changed would be a grave understatement. With my housing needs stabilized, I began to reassemble the other fractured pieces of my life, but the lure of easy money would once again tempt me. Family members and friends threw me a housewarming party. It was really nice—a lot of people I hadn't seen in decades, including those from a past I didn't want to visit. As the night went on, most of my biological family left. When the others began to leave, a group of old friends who were still earning money the same ole way had "gifts" for me. When it was all said and done with, I had been given fifty-nine grams, a little over two ounces (a street value of about $4,500). I was presented with a huge dilemma.

As I cleaned my apartment from the party, I came across things that gave me pause, which belonged to my mother, and pictures of my son, grandfather, and friends. And then it dawned on me the promise I had made back in my tent. This stroke of good fortune was a test of faith! With that epiphany I had only one choice. I flushed the dope down the toilet and prayed. I have been sober ever since. It was from then on that I found a different path, one I never would have found had I not struggled in Silicon Valley. I became an advocate for the homeless, the addicted, the forgotten about. I try to assist those who are in the dire straits I emerged from.

EDWARD NIETO

BLOOD-SUCKING MACHINES

At the bottom is where we all must start, unless you know someone. Working a job that pays well sounds great, but you most likely won't get it without starting at a temp agency. Sometimes you are the youngest person there, with a trainer who won't show you how to do the job right because they're afraid you will take their job. After a while, you hate what you're doing, but you've got bills to pay. So you just think about the money you get per hour.

I have worked at least ten different jobs in Silicon Valley. Most recently I worked as a regent at Abbott Labs, running cardboard into the machine making boxes, labeling the boxes, making sure each box had the right bar code, then packaging human blood and other chemicals into the boxes. I would put these plastic tubes—called bladders—inside the box and then put them into another machine—the sealer—to seal the box. And that's what we did all day, or until they moved our positions, or until lunch.

I would start my day by opening up my locker and getting out my hairnet and smock. Then I would go into the next room to get my earplugs and rubber gloves. People who had beards or mustaches had to wear beard nets too. We replaced the night shift people. We were the first shift of the day, and we had to be there at 6:00 a.m.

I would work ten or eleven hours a day. I had Sundays off. A lot of times, I didn't know what we were packing, unless it was cyanide (they let us know if we were dealing with poison). Even if we were, they didn't give us any safety equipment. They just told us so we wouldn't touch any chemicals. We also placed a lot of human blood.

When I first started, I felt weird about working with blood. They told me that I would be working with medical supplies, but they never told me that I would be working with blood. I mean, it was blood.

Before I started as a regent, I worked in the warehouse where they got the blood in little white boxes, and we would pack it in ice and styrofoam all day. It was sent out to other cities, other countries; there was always a huge order for blood.

The company would sometimes ask us to sell our blood. The first time they pay you $14. The second time, it's a little more. The third time it's $25. You can do it three times and then you have to wait a while. Whoever needed money that day would sell their blood. They are always in need of blood.

I did it because there was a news story on how hospitals were short of blood these days. After I gave blood, I told this to one of the nurses and she kind of laughed. She told me that they didn't use the blood for humans, they used it to test the medical instruments!

I did the job for three months and thought I would be hired for a permanent job. But I was let go, on the same morning that my grandmother died. My assignment had "ended." It was an all-right job.

K. S.

TURNING METAL INTO MONEY

Aluminum, tin, brass, metal, and copper—they all scream out money to me. With the tough economy, I have to pull pennies out of the dirt. I'm what some call a "scrapper." I pick up scraps of recyclable items left out by companies or demolition sites, and recycle them. It might sound easy, but the art of scrapping is something that takes time and talent to master. It's like a game where time can be your best friend or your worst enemy.

To scrap, you usually need a truck or a van to haul off the large items. Then you take them to a recycling garage where prices are posted daily with how much your items are worth. Usually, there is a scale for you to weigh your items on. Basically, any item is worth money as long as long as it doesn't stick to a magnet. It can be almost anything: wire, pipes, tables, computers, large complicated equipment that can be taken apart and stripped for cash.

But don't get it confused. I don't steal or vandalize; I take and keep the environment clean. But some folks have given scrappers a bad name by destroying property in order to get a bit of copper wire. Lately, there has been a lot of controversy behind scrapping, and law enforcement has really made an effort to crack down on scrappers. They have been advertising their efforts in the news and setting up traps. When I say traps, I mean leaving small amounts of shiny aluminum for the scrapper to take and BOOM— caught on camera.

In scrapping, timing is everything, disguise is what will make you invincible, and knowing where to cash in and slang your scraps to the fullest is what makes you your green. I was taught how to be scrapper by an O.G. He showed me all the hot spots and how to scout. Since I'm a girl, it's easier for me to get better prices because it's rare for a pretty woman to step out and negotiate a man's work. On an average five-hour day I can make anywhere from $200 to $600. With that kind of income why go back to a nine-to-five? The money is fast, easy, but if you don't know what you're doing, it could lead to a felony.

Companies and construction sites have grown frustrated with scrappers, because some scrappers don't have respect. They leave a mess, take new products, and destroy sites that then take several months to rebuild from the damage made. You got to respect other people's time, because if not, time won't be by your side. Another reason companies are upset is because scrap piles are worth a good amount of money. Therefore, they don't want scrappers to come up off their profitable items. As a scrapper, I carefully select what I take. I prefer to take items that are really just scrap, not brand-new items that will be used. My favorite items to come across are copper and aluminum. Other metal like tin and brass aren't worth as much. For copper you can get up to

$2.50 per pound. So if I have two hundred pounds, you're looking at an easy $500 just to pick up money from the floor and transport it.

The scrapper lifestyle can be addicting. It hurts to go back to paying taxes and making an "honest buck" when what I do is hard work, and at the end of the day I feel like it's good money earned. I love scrapping. It's fun and it gets my heart racing. It also helps the environment and pays the bills. Always remember, when you're making fast money, give some back to the community, whether it be to a homeless person on the street or your little sister wanting you to buy her an ice cream. Never be greedy with free money. I have turned down jobs, and my hustle notch, because scrapping isn't my career. I just use it as a temp job when necessary.

To all you prospecting scrappers, don't go on a rampage and take any metal piece in sight. Weigh out the scene, in and out, be quick, and always bless and give thanks to the business you take from. My palms are itchy; I think it's time to clean my environment.

ALI RAHNOMA

CONFESSIONS OF A CELL PHONE SALESMAN

"Hey, you want to buy a cell phone?" a voice snapped at
me as I walked through the mall. "Nah dog, I'm straight," I
chuckled back at him. His half smirk, wrinkled dress shirt,
and crooked nametag reminded me of my own countless hours at various retail stores selling phones. "What if
I could get you any cell phone for free?" he continued. I
laughed out loud as I called his bluff. I've been making a
living as a salesperson since I was thirteen. I've gone from
hustling tables for tips at my parents' Afghani restaurant to
hustling my friends and family with cell phones. Regardless of what I sold, the game always stays the same: You
either play or get played.

"Playing it" means coming up with creative new ways of
making money. Sometimes it means ripping off the company,
other times it means ripping off customers, and often it means
doing both. I can vividly recall my recent days as a salesperson
at a ma-and-pa cell phone store in Milpitas. To supplement our

income, a coworker and I opened up a chop shop in the back of our store. We used to get old phones, fix 'em up, and sell them as our special "free phones" to unsuspecting customers. We'd send the phones we couldn't fix back to the manufacturers to get repaired. When they came back brand new from repair, we'd use them for a week, and then we'd sell them back to customers for cash. Easiest money I ever made. Sometimes, it took a while for rookie sellers to learn the rules of the game. They would run up to me and say, "Yo Ali! I'm sweating bullets. This crazy customer wants to talk to the manager!" I would go outside and introduce myself as the manager. I would pretend that I was interested in hearing what the customer would say, then calmly apologize for the problem. Most of the time, customers don't want their actual problems to be fixed. They just want to find someone that cares about what they have to say. That's how I learned to smile and nod. A real lesson a customer should know is, "Never mess with someone that has all your personal information on file." One of my homies that I used to work with took that motto to heart when he was confronted with a particularly angry customer.

The customer was a rich techie who would drive his Porsche up to the store once a week and yell at us for an hour about complaints he had with his service. Whenever he finished with his routine, he would leave the store belching out racial slurs. One day, my creative colleague decided to activate three new lines at the expense of the customer. He made sure that everything was clean so the transaction could not be traced back to our store. During that weekend, my friend drove up to the city and dropped off the phones at three different random locations in the ghetto, hoping that someone would pick 'em up and use the phones to call their relatives across the country. A month went by without a trace of the customer. We later found out that his credit rating was so shot that he wouldn't even qualify for a Chuck E. Cheese credit card.

The way I see it, slanging cell phones has its perks. You gotta learn to hook yourself up when sales are low. One of the perks is that everyone wants a cell phone, and, if you're cool with the clientele and you know how to manipulate the company's system to your advantage, then you've got it made.

I had a regular customer who would come into the cell shop twice a month on cue. Each time he would bring in a different friend. I would cheerfully call his name just as soon as he took his first step into the store. His eyes would light up with a smile every time he heard his name. He loved being acknowledged, especially in front of his friends. As soon as he heard his name he would randomly grab something off the shelf and skip towards my register, shoot the shit for a minute and then hand over his credit card without looking at what he was buying. I'd give him a 10 percent discount, enough to barely cover the tax. In the three years that I worked at that particular store, I must have sold him the same product at least five times. That's when I learned that keeping the customer smiling meant that my money would start piling.

During downtime (usually during the months following Christmas), most sales people generally quit because of the lack of customer flow. But an experienced hustler knows that is the best time to tap into the network and hook yourself up. That usually means relying on the local underground economy to keep things rolling. This works best if you work at a large shopping mall. A cell phone in trade for some new name-brand clothes is not a bad idea, for example. But that's just my preference. I've seen cell phones being traded for drugs, sex, PlayStation 2s, and all kinds of weird services. The funniest trade I've seen was a $200 cell phone for a bag of weed, some flowers, and an industrial-strength air filter. You figure that one out.

Considering my encounter at the mall, I realized that everyone participates in the game. That's just the way things are run when people are hustling to make ends. One of the most important things that I've learned from the game is that it can eat you up if you don't have a game plan. The constant desire to want a fatter paycheck can eventual-ly distract you from the bigger game: life. Despite that, at one time, I was making money hand over fist compared to other young folks in the Silicon Valley. Of course, I was also spending it all from paycheck to paycheck. I was seeing the cell phone companies slash our commissions and change strategies to adjust to the needs of the economy.

Nobody wants to be a hustler for life. Even the biggest hustlers want to move up and out of the game. Hustling means survival, and generally being in survival mode will take its toll on a restless mind. I retired from the game because I was beginning to feel burned out, and I realized that I did not have any real dreams in life. I'd like to think that I made sure that I did not play myself, let alone get played by the game. Hanging up the jersey wasn't easy, but everyone has to leave the game at some point.

CECILIA CHAVEZ

CLEANING UP THE VALLEY UNDER THE TABLE

As an undocumented worker, I have been able to sustain myself by holding jobs where I was paid under the table. I started cleaning houses at the age of nineteen. My older sister was already working in the house-cleaning business and she offered me a job. I took the job because as a college student I would be able to have a flexible schedule that would allow me to attend classes.

When I started working with my sister, her business consisted of about ten to twelve houses. Her clientele was primarily families of Indian decent. After I joined, we were able to take on more houses because we were able to clean faster. Eventually our business grew to the point that we needed to hire an extra person to help us with the houses. After two years of house cleaning our clientele had grown to more than thirty houses. We worked six days a week from eight in the morning to eight at night, sometimes even later. Cleaning houses is not an easy job.

Each day varied. We would usually clean between five to eight houses a day depending on how much time each house took. The hardest part of our job was its physical requirements. Your body never gets used to the up and downs throughout the day. I had to bend under beds to clean the floors, get on top of furniture to reach high drawers, bend again to clean the bathrooms, climb stairs, bend again to pick up the mess in the living room, move furniture around, and repeat everything again in the next house. The worst part for me was when I got home and was able to rest and lie on my bed and not be able to move because of my sciatica pain. The only thing that I could do to alleviate my pain was to lie there and go to sleep.

When I first started, I was eager to work. One of my qualities is that I am a neat freak and like to keep everything clean. Stepping into a dirty house was actually exciting for me because I was able to make a visible difference and show off my skills.

A common thing that most people don't know about the house-cleaning business is how clients try to set up traps to try to catch you stealing. That would happen to us when we first got hired to clean a house. The family would intentionally leave valuables in open spaces. My sister, our coworker, and I already knew that those were setups the owners would leave to see if we would take anything from their house. But our parents raised us right. So the money and electronics that were left lying around in the houses that I cleaned never tempted me. I knew that I would lose more than what I would gain.

The best way to see wealth disparity in Silicon Valley is to be a house cleaner. Sometimes I couldn't even tell I lived in the same region as the people whose homes I

cleaned. I live in East San Jose in a house that is too small
for my family and doesn't have the commodities that other
houses have. But when we left each morning to work, we
drove to the nicest parts of Silicon Valley to clean houses
that we will never be able to afford. The biggest house that
we cleaned was in Los Altos, California. That mansion was
home to a family of four where both parents were engineers from India. The house was enormous; it had four
huge bedrooms, two kitchens, two living rooms, a library,
a music room, two office spaces with diplomas hanging
from every wall, and an enormous backyard that seemed
big enough to raise cattle in, with a swimming pool. The
master bedroom was bigger than the one-bedroom apartment in which my sister lives with her husband and two
children. To clean the entire house it would take us about
five hours.

The first time I walked into the house I was amazed. I
couldn't believe that houses like this existed in real life. It
looked like a house out of a reality show. Every time we
would go clean this particular house, I always imagined
what it would be like to live in a house like this, or be able
to afford it. When I would change the bed sheets, I would
wonder if the owners ever thought about the people who
changed their sheets while lying on their crisp, clean bed. I
wondered if they if they ever imagined themselves in our
shoes. I certainly imagined myself in theirs.

But despite the enormous wealth, and the amount of
work we did to clean that house, we were not compensated
right. We were only paid two hundred dollars for the work
that we did. For us that amount had to be sufficient because we did not know how to negotiate or let the family
know that we needed to be paid more in fear of losing a

client. After asking around to other cleaning companies and comparing prices, we learned that for this house we needed to be paid at least four hundred fifty dollars.

But that was usual for us. We were never paid what we deserved; we were paid what the family thought we should be paid. Because we were undocumented workers we didn't want to upset the families, so we accepted what they offered.

I stopped cleaning houses after I got offered a babysitting position from one of the families that we cleaned for. Although it was a difficult decision to leave my sister, I had to make the decision that was best for me. My body needed a rest from the physical requirements of house cleaning. After I left our company, work was too much for my sister and our coworker. They could not continue on with the same work pace because it was too much work for them to handle. Slowly, my sister started letting go of some of the clients that she had gathered. She stayed only with the houses that paid better. Although it is a demanding job, she continues to do it because it is her way to sustain herself. I continue to help her once in a while when I have the time and energy, because despite my education—I've graduated college since then—it is something that I enjoy doing.

TAMEEKA BENNETT

LYFTING UP MY INCOME

One day I was looking at my paystub—comparing it to my incoming bills—and I said, "Yeah, I need a second job to help make some of these ends meet!" Enter Lyft.

I needed something that wasn't time-consuming, that would allow me to set my own hours and offer me a livable, albeit part-time, wage. And so started my journey with Lyft. The process actually went a lot quicker than I expected, which was a little surprising to me. I signed up online, submitted my insurance, pictures of my car, and a copy of my driver's license. After a background check, I received an invitation to drive with Lyft. In order to begin driving, I had to meet with a driving mentor, have them check my car out, and go out driving with them. It kind of felt like I was eighteen again, trying to pass the behind-the-wheel DMV test. I had a woman, she was pretty cool. She's been with the company since the beginning. She's one of the original drivers, and hearing her story was telling of the many folks that turn to companies like Lyft. Her story was reminiscent of my own reasons for joining. She was a single mom living in Silicon Valley trying to make ends meet when she discovered Lyft. She loved that she could set her own hours because that left her time to be there for her kids. After a few months with Lyft she decided to be a full-time driver and hasn't looked back since. I passed the test, with flying colors, may I add, and started driving with Lyft within the next couple of days.

I was pretty nervous but excited at the same time on my first day, because I love meeting new people. So the idea of having strange people sit in my car while I drove them to their destinations didn't freak me out.

As a driver, the time you spend with a passenger can go a few different ways. You can have on some jamming tunes that do all the talking for you, you can allow them to get lost in the tiny world of their smartphones, or you can strike up a conversation. I always choose the convo. I can't help it, I like to talk.

Well, I've met some interesting Lyft patrons in my driving adventures, but the most irritating had to be this one guy that I picked up from a tech company that shall remain nameless. I'd picked him up after what he told me was a long day at work, so maybe that contributed to his not-so-great demeanor, but eh, who knows.

So I'm sitting in the car with this guy hoping that whatever small conversation we have will help turn his day around even the tiniest bit. Well, that didn't really happen. We spent the car ride talking about why I was using the wrong navigation software and who I should be using, a certain actress's rack (yes, he talked to me about boobs... like, for real?), how everyone hated him in high school because he was smart, and some other crap. After he got out of the car, I literally sat there for like a full minute thinking, *Who is this guy and what could have possibly happened in his life to make him that way?* After I took my minute, I prayed against any bad juju that may have been lingering and went on to my next pickup, who was by far a cheerier passenger.

As of this exact moment, I am on a hiatus from driving for Lyft. I'm going to pick it back up in the next month. Yeah, there were (and still are) a couple of things that I'd like for the Lyft community to address, like safety for drivers. We have background checks run on us for security reasons; I'd like for the same thing to happen for the passengers. I've had to take people to some "poorly lit" places, and I remember thinking, *What's going to be your attack plan if this dude turns out to be crazy?* I even had a passenger ask me if I felt safe as a driver. She happened to be a professional security assessor and thought it strange that as drivers we aren't better protected. I agreed.

SAID FARAH

DRIVING A RICKSHAW IN SAN JOSE

Being a struggling young artist in a foreign city with no family to speak of is not easy. So when I found myself facing destitute poverty and still sending out resumes and waiting for callbacks, I decided to think creatively.

After fielding random ads on Craigslist for temporary work, I found a pretty enticing offer. The concept would be to pedal my ass off for tips on a three-hundred-pound pedicab rickshaw, with drunken passengers on board. At first, it seemed like there were no drawbacks. I'd be able to get into good shape (which is hard to do as a half-pack-a-day smoker) while being entertained by tipsy tippers. More importantly, I could do it consistently and mold some semblance of a life around my hard-earned crumpled dollar bills. I really had nothing to lose, so the least I could do was give it a try.

I started riding around for menial tips the weekend following my run-in with the company. At first, the work was grueling to say the least. I was leading an extremely sedentary lifestyle at the time. I was eating once a day, if that, so it was beneficial for me to use as little energy as possible to stretch out my meals. I usually worked three-hour shifts at first to build up my leg strength and quickly switched over to seven-hour marathons for more pocket money. At one point I was sleeping in my car but quickly found a situation that would allow me to stay rent-free for half a month while I saved up tip money. Being hopped up on Red Bull and hauling heavy human cargo for extended periods of time was quickly getting tiresome. I was coming home zombified every night and eating more than a pro-football offensive lineman every day. But I quickly fell in love with my newfound source of income.

Now, I'll be the first one to tell you: the perks are what make me look forward to my pedicab hustle. Of course people want to drink with you, among other things, but it doesn't stop there. Two words, my friends: eye candy. I guess I'm no more superficial than the next Joe in line, but it's kind of hard to avoid scantily clad women with plenty of social lubricant in their systems. Does it help that the carts light up like a Christmas tree at night? Hell no, but you won't find me complaining. The other night four gorgeous women hopped onto my cart, I mean way out-of-my-league type of girls. They had just gotten out of the club and needed a ride back to their car. The trip shouldn't have taken more than five minutes, but I was forced to circle around for no apparent reason while deflecting their constant barrage of propositions. I'm telling you folks, I got no less than a thorough massage that night. If I weren't black, I probably would have been blushing. Afterwards, I had to pinch myself to make sure I didn't doze off in front of some bar and start fantasizing on the clock.

One weekend I was offered to go down to Carmel for a special event, with the chance of running into celebrities and getting guaranteed money as well. It was slow going at first, but I knew it would be a good night when my first tip turned out to be a $20. Halfway through the night, after finally recovering from the shock at the slew of trophy girlfriends and wives outside every establishment, I ran into some former NFL players. One of them had won three Super Bowls with the 49ers in the eighties and ended up giving me personal advice for close to an hour. What shocked me was how humble and down-to-earth he was, considering the amount of success he enjoyed in his life. At this point, I was feeling pretty good. There was a steady flow of generous tips, and I had gotten fully acclimated to the treacherous hills of the downtown strip. While I was killing time outside of a bar with the world's most interesting limo driver, my buddy lets me know that Bill Murray just got on the dance floor.

I decided to entertain the notion and strolled casually to one of the modestly sized windows. There he was, with that trademark nonchalant look glazed over his face while diving into a late-night snack. To my surprise, Bill Murray ordered another drink and proceeded to throw down on the dance floor. That made my night, hands down. The best part was, I must have counted my money at least eight times on the way back to San Jose. I could actually afford a car wash. This rickshaw thing was really starting to pay off.

So I guess I've been doing this for a month and a half now, and I've somehow managed to gain fifteen pounds while losing body fat. The only downside is the inconsistency of cash flow. No two nights are the same, and you never know what you'll run into. I've seen people stage dive off of moving rickshaws to avoid paying a nonmandatory tip. I've had countless frat-type dudes asked me ever so eloquently if they could touch my "wang!" I've made friends with people that I've given rides to, which is a huge plus for someone who's only been in the area for two months. Overall, I can't complain. The rickshaw has been good to my body and me, so I don't see myself stopping anytime soon. The next time you see an ordinary-looking biker hauling around overweight drunkards up Himalayan inclines while chugging copious amounts of Red Bull, make sure you get a ride with them on their way back down the hill. And, please: tip your driver! You don't know what we go through.

MIKE SANDOVAL

RIDING THE ROW

I have a dream job: I park cars for a living. I landed a valet
job in Santana Row, the high-end shopping mall in San
Jose. Getting the job was easy. In the interview, the guy
never even asked to see my driver's license—he was only
looking to see if I got along with him. And on that note,
he hired me. I was blown away from the first day. I drove
cars that I thought I would never drive in my life, like a
Porsche 911. It took me a while to start the Porsche because the ignition was on the left side instead of the right. I
hated looking like I wasn't familiar with the expensive cars.
Every time I jumped into a car, I tried to look confident,
even as I was telling myself, "Please don't stall."

After a while on the job, I noticed that everyone who
got hired at the Santana Row spot was a friend of the head
valet. It was like some underground community. And that
was okay by me—I needed the job.

On one of the first weekends there, I was just chillin' out
at the front desk and an orange ragtop bug pulled up to the
curb. It was a San Francisco 49er football receiver who was
in the news. During the off-season, he lived at the Santana
Row apartments. He told me that he was leaving for the
weekend and needed to keep his car safe. I said, Okay, I'll

pull in the sweet spot for you." So the weekend went on and that Sunday I decided to take a cruise in his car, just around Santana Row.

While driving it, I was bumping his $60,000 sound system. It was bad. When he came back from wherever he had gone, he found me and told me another 49er player had seen his car cruising around.

I didn't know what to say. I knew you couldn't miss his car because it had his football number stitched in the seats. But I answered him without missing a beat, "I just moved your car from here to here," pointing to different sides of the lot. He said, "Okay, that's cool, that's cool," and tipped me twenty-five bucks.

As it turned out, weekends are the best time to be a valet at Santana Row, as that's when all the drunken tippers come out of the bars. If I was a cop, I would be handing out DUIs like coupons. You can learn a lot from seeing what's hanging on people's visors or layin' on their passenger seats. People can be a lot different than they appear— like some old white lady will be listening to salsa and hip-hop. And I've seen panties and porno tapes in cars you would never expect.

But Santana Row weekend partiers also bring their drama. One Saturday night, a lady, obviously drunk, pulled up and threw the keys at Pablo, one of my coworkers who barely speaks English. She told him he could have the car! He apparently understood and left with the car that night. The next day the same lady showed up and asked where her car was. She had only a vague recollection of the night before. She remembered throwing the keys at Pablo. So we called him and told him what happened. He told us that she gave him the car. The boss told the lady what Pablo had said. She was upset and told the boss that she'd had too much to drink, but that she didn't think anyone, especially a valet, would take advantage of that fact. In the end she just wanted her car back. When Pablo obliged, she was grateful and didn't press charges. Interestingly, Pablo was let go shortly after that incident.

All in all, being a parking valet in the most exclusive part of San Jose can be fun and full of surprises. I would recommend it to anyone with an adventurous spirit.

EDWARD NIETO

THE SIGN

It's funny doing this work—standing on the street dancing on a street corner with a sign while jamming to tunes and waving at people driving by. Some people think of it as degrading, but in reality it's really uplifting.

Some people may have jobs where they can listen to music while on the clock, but how many people are able to dance at work? I can listen to my music without any complaints—it's just me out there, with my favorite artists keeping me going. I've worked plenty of other jobs, some for the same pay rate, but none of them gave me the freedom this job does. When I think of those jobs, like in fast food or the factories, I thank God I'm a sign waver.

When I've worked at fast food, it felt like a military job with the shift manager yelling in your ear. All day, all I heard was him barking about getting more meat restocked—"More meat! More meat, More meat!" Plus, those jobs usually require you to do multiple jobs for one paycheck. You have to stock, flip patties, clean restrooms, take out the trash, basically everything. The factory jobs that I once had carried the same pressure, bosses always on your back, plus the risk of injury from the equipment. But being a sign holder is just that: the job description is

self-explanatory. There is no dish duty, no cleaning surprises in the restroom, no risk of getting hurt on the job.

I've found what a lot of people said was impossible: a
stress-free job.

Those other jobs were rough, and here I am getting paid
to be in my own groove publicly. It's funny when I see
someone I know drive by, and they're in their work out-
fits, and I'm in mine—which can be a bit unusual for most
work attire. Sometimes I'm dressed like the Statue of Liber-
ty, or something that looks like I'm the missing member of
the music band The Village People. Besides a good costume
that is eye-catching, the most important thing to prepare for
this job is your playlist. While some other jobs frown upon
employees listening to music while they work, it's almost
mandatory for this job. You can't forget the sunscreen while
dancing in the sun. And you must have your earphones in
and an MP3 player because you have to keep moving. It's
the only way to keep moving and to have a great time. If
you're not happy, the people in the street are not happy.

But if you're happy, everyone around you is joyful and
waving back at you. The real success is when you see people open up their eyes and see your sign, and then you see
them in the office—it's a good feeling of a job well done.
So when you drive by a sign waver dancing, twirling the
sign, make sure to smile, wave, or honk—they'll appreciate it. Truth be told, these jobs aren't going to last forever.
Like many other great American jobs, even being a sign
waver is becoming automated. These days you might see a
manikin-looking thing carrying a sign, which is sad for all
kinds of reasons. But until I get replaced by one of them,
you can find me on a local street corner, sunscreen on,
earphones in, getting paid to be in my groove.

ANDREW BIGELOW

GOLD OUT WEST

*I walked from East San Jose to the campuses of Apple and Google
in Mountain View with marchers on fixed income. It was about
about thirty miles. We wanted to show the tech giants who else
lived in Silicon Valley. It was called the March to Heal the Valley.
This is the poem I wrote and recited as we marched.*

They say my city streets used to be paved in gold
The home of where they used to manufacture Silicon
Big city, still easy to feel alone
Surrounded by wealth, left to wonder why it's not my own
They make the iPhones across town from battle zones
Where they think government aid can fix a broken home
I've never known what it's like to live that life
But I got folks across the street starting fights in broad
    daylight
Little kids play on they bikes
Down the street from the house that slangs white, main-
    taining through the night
They say my city streets used to be paved in gold
Now workers laid off and homes foreclosed
This was the illest rhyme I ever wrote
Cause I wrote it in a state of hope

Politicians' speeches like a bad joke
Talking about the people but they really about that cash
    flow
Had to let your ass know

Now you know where I'm from

There's gold out West but not enough for everybody
So somebody's scheming somebody
There's gold out West but not enough for everybody
So somebody's robbing somebody

This the place where folks make bills like Bill Gates
Move into private communities and build gates
The land where corporate don't cooperate
Hiding money in Swiss banks
Others in Section 8 with they bills late
Mansions in the hills, needles in these gutters
We don't see the bigger picture, end up fighting with each
    other
I'm in my front yard lighting herb
Hearing pop pop, is it gun shots or fireworks
They say my city streets used to be paved in gold
Now people living in tents with nowhere to go
It's getting intense, City putting more cops on patrol
Trying to get these violent streets controlled
But really, they only locking up our family and our folks
And I'm stressing yo, Mom's receiving eviction notices
While CEOs getting bonuses

They give a fuck bout us, they about that cash flow
Now your ass know

JEAN MELESAINE

NEW BOOTY AT WALMART

For my entire life I've never had a real job. I always made
money in other ways, which made me feel like the biggest
hypocrite ever. I always tried to balance out the good and
bad: the good being education, and the bad being crime.
I'd always emphasize to my younger brothers how important it was to be educated and live good, which was bullshit
because I was still doing illegal things. Those illegal things
caught on to my younger brother, and I played catalyst
because I was there supporting him. To change both our
thinking, I thought I should show him the concept of hard
work and finally decided to enter the economy not as a
consumer. Plus, I was broke.

Entering the economy wasn't as easy as I thought
because I had no prior experience except in community
organizing and nonprofit work. Most corporations could
really give a damn about you organizing a community. I
went to Walmart because a friend told me they hire anyone. I soon found out she was right because my brother
and I both got jobs even though we both had criminal
backgrounds, probably had the worst interviews in Walmart

history, and dressed like shit. We were hired within two days. Because this was our first job ever, my brother and I were mad happy, my brother especially. He would chant frequently and proudly "I got a job" or "I have to go to work," even wearing his vest and nametag when it wasn't necessary. Getting hired was only a short prelude though, as we soon found out. This job would completely change my thinking about entering the economy. The training was mostly done on computers, tests that could easily be passed by an elementary student. They even had a video that bad-mouthed unions.

Our first day was fairly easy. We mistakenly came to work on the dayshift, which was very slow. We walked around the store for six hours just touching things and pretending to look concerned at merchandise. The following day we soon realized what this corporation was all about. We came in as nightshift stockers and were introduced as new booties at the beginning meeting that was held every day for stock purposes. At the end of the meeting, not that it matters, but there's a really corny cheer they have the employees do called "The Squiggly Wiggly," which I thought was ridicule at its height.

Our first night, they worked us to see where we were at with our health. Being that it was our first real job and our first night, my brother and I purposely kissed ass like mad crazy. We worked fast and clean, and we did it with the most kiss-ass smiles. That was the wrong thing to do because the supervisors saw how fast and hard we were able to work, and placed us in our own departments. My brother was placed in Chemicals and Paper Goods, and I was placed in Pets. Our departments were clear across from each other, and we always talked smack to each other. They would have me stack fifty-pound bags of dog food while I watched my brother stack napkins. The smell in my department was unbearable, but I got used to it.

I usually finished before my brother, so I always went to help him in his department because by this time he was going crazy over his job. Our supervisor gave him a warning that he was working too slow, and if he couldn't do the job fast enough, they could easily find someone who could. These kinds of threats were made often to all employees who didn't meet their working standards. I never really got mad because I knew that they were only threatening us because they were given threats from someone in a higher position. Walmart is pretty much like society: low positions have to screw over other low positions to move up, only with the result of being nowhere near the top. After a month of getting treated horribly, my brother and I became the worst employees ever. This is why you should never hire former criminals who hate to get messed over. We completely lost all our morals of trying to be good. We'd place merchandise wherever we pleased, ate and drank whatever we wanted, started stealing DVDs and watching them on our breaks in my car. I slept on the job because I found out that my department was actually the best department in that store. Pets was the only department without security cameras. The stock slots for the pet food bags could easily hide you if you were to lay there. Bagged pet food is a hard flat surface, but the good thing about that is they make very large bedding for pets that was more comfortable than a lot of human bedding. I'd make my own bed, and either read or fall asleep. I even started to read tabloids and got deep into that "Brangelina" bullshit. After four months, my brother finally quit. He couldn't handle the work no more because they completely took advantage of us for being young new booties.

When school began, my schedule interfered with everything. I would go to work from 10:00 p.m. to 8:00 a.m. and head straight to school from 9:00 a.m. to 2:00 p.m. Redbull and I became very close during this time. For two more months, I tried to keep my job and go to school to be somewhat of a good example. That soon changed after I got a new supervisor named Daniel. Daniel was an asshole. If they had a new visual dictionary, his picture would be found next to it. I always sensed he had something against me because he always boasted about me being a student who worked. He would have me finish my department before everyone else just so I could pick up after the other employees who were slacking.

My checks were usually spent on my family. Since my work schedule conflicted with my weekends, I had no social life. There was no point of me spending money. Still going to school, I saw a documentary in class on the Walmart corporation called *The Cost of the Low Price*. My professor hated Walmart with a passion (she had no idea I worked there). After watching half of the documentary, I quit, but I quit in the right way. I waited to leave on a day they needed mad help. Looking back and thinking this out thoroughly, I see that Walmart was the worst place to begin my working life. It made me not want a nine-to-five job ever again.

DAZEL VALADEZ

EULOGY TO MY MALL JOB

Seems like everywhere you turn in Silicon Valley, old-school stores we grew up with are closing down. They have been left in the wake of online shopping. In fact, I was passing by my old store to say "What's up" to some former coworkers, and when I looked over to my right, the store I spent years working at was closed down.

I had worked for that company for six long years and have so many memories that gave me some of the happiest times of my life. It was my first real job, a little gig that I had right after I got out of high school. I didn't expect to work there for so long—it was just supposed to be something temporary while I was going to college. I came to discover that I loved being at that store because of its clientele.

It was one of the only legit stores on the East Side that dressed up the homies and had the cutest things for the girls. It catered to all styles and types: the gangstas, the pretty boys, the "hoochies," rockers, basically any look. It was where everyone could shop and find something they liked. What made it an even a better place to be was we always had the newest and dopest songs playing in the store, so besides working, I was always dancing everywhere. Even one day when my manager said to me, "This is not the club!" Having me dance was half the reason why people came in to the store.

I remember helping a teenage girl looking for some fits, and her mom was shopping with her. She was like "Your makeup is hella bad, ey!" While thanking her for the compliment I began to notice she had a tall can wrapped up in a brown paper bag that was sitting in a cup holder on top of a stroller! Then there was my favorite, "Cologne Man." He was an older guy who came in at least three times a week and he would come and talk with me and tell me about his daughters. He would always spray himself with the all of the cologne testers we had and afterwards would walk around feeling like a million bucks. It didn't matter what he would try to buy, he always had to try to bargain to get everything dirt cheap! I always wanted to tell that guy that this was the Eastridge Mall, not the flea market!

From working at the store, I learned many things about how to talk to people and how to sell. When your customers start to know you and feel comfortable with you, they start to build a better relationship with you and trust you more. Once that trust is built they are more open to your opinions and value your advice a lot better. My selling skills improved because I learned how to read people and be able to tell what kind of style they got just by looking at what they were already wearing. From doing so, I learned how to sell them items that they actually liked and not just try to make the store that extra buck.

My store closed down because they claimed bankruptcy. They started closing a few stores here and there about six years ago, and as the years went by, more and more stores started closing. Right after Christmas I guess they finally kicked the bucket and filed. I have worked at other malls and in other stores and I've seen plenty of stores start to close down. I guess with this economy you would be lucky if any company survives as shopping goes online. After going from an associate to an assistant manager, I think I am done working in retail! It was a nice experience and it was something that was helping me get through college, but I think I have outgrown that business. No matter where I end up, I don't think any other job can compare. It was by far the best job I ever had, and I'm going miss the crazy characters from East Side San Jose that I met.

MIKE NICE

SOMETIMES IT'S FUN, SOMETIMES THERE'S BLOOD

I'm not a violent man. I like to see people have a good time, but I don't like to see them stupid-drunk, acting like clowns, and classless. I like to enjoy a woman's beauty, but I don't like to have strange, drunkass women fondle me as if I were their personal rubbing pole. I like to dress up, but I don't like having to stand in the same spot for four hours, wearing a stiff two-piece that makes me feel like a bodyguard. I like protecting people and breaking up fights, but I don't like being an asshole, which is what people are used to when you do what I do. I am a bouncer at an upscale San Jose nightclub. Sometimes it's fun. Sometimes there's blood.

I actually love the work. I don't see myself as a bouncer, but more like a protector. My job is to protect the property, the fun environment, and people's safety. I like breaking up fights, pulling dudes away from conflict, talking to them, and making sure the beef is put on ice.

It's not always that simple, however. Besides being the main producer of revenues and an item of desire for clients, liquor makes people think they can do things that they really can't—like fight. Sometimes, as the night progresses and a person gradually gets drunk, you can see how they can become a walking misunderstanding, such as in the situation of a girl being harassed, two drunk dudes accidentally bumping shoulders, and the ensuing confrontation in the end.

I try to see the humor in it. They'll show up between 10:30 and 11:30, get a drink at the bar, stroll around looking at themselves in the mirrors, walk right by me as if I were the wall (few folks greet bouncers, probably because most see us as cops), get another drink.... The walk gets a small limp, and eyes get a little crooked. Then more people arrive, and the same early customers are stumbling by midnight. By 12:30, you can smell the liquor on the crowd, begin noticing which clients might have an episode of stupid drunkenness and get into a fight.

I always check in with people—men and women that carry a hard "I'll fuck you up" look on their face. "How you feeling?" I'll ask. "I'm cool," is the most common response. "All right. Just want to make sure because you look like you want to hurt somebody." They almost always laugh, and it's most times a dude. "Let me know if you feel like doing anything dumb," are my last words. People will let me know if they think a fight is about to erupt or if a girl is being harassed. I am always as professional and polite as possible, never pushing, commanding, or being an asshole. People respond well to this approach, and it feels good when people tell me that I'm a cool bouncer, different from the others.

But sometimes the fight just happens, and we bouncers are the first to jump in, pulling people away, dodging punches (or not, and taking a few), calming people down, and getting the fighters outside. Most dudes don't know how to fight well and are just drunk, so they don't care about who they hit or if they are hit. The liquor numbs the senses, which includes the senses of maturity and honor. Sometimes girls get hurt. The chaos explodes, and all of a sudden, a body collides into you, and you hear bones and flesh getting beat. A lot of times, the looks on people's faces are worse than any punch they might throw. It's drunken, frustrated, hateful, and exploited rage. I know my peers are overworked and underpaid, exhausted by the time the weekend comes around. They want a female body to lay with, but miss most of the time because they aren't rich or good-looking enough to get that one-night stand they're craving. To make it all worse, they can't see themselves from outside of themselves. So they don't understand why they want to get drunk and fight, and wake up feeling shitty the next day. My job as a bouncer is to push the violence outside the club, which I don't like to do because I would rather help end it. I hate seeing my peers make war with each other, because nowadays, fistfights become knife and gunfights too fast.

When I'm not telling people to clear the hall or telling some drunk pervert to stop hanging on the women, I'm talking to people, telling them where the bathroom and smoking areas are and talking to girls. That's one thing about being a bouncer in a suit: the ladies like it. Part of me enjoys the attention; another part doesn't care and just wants the night to end. To be sure, I hate when drunk, slurring women with hot liquor breath come and try to feel me up. Somebody must have told them that that's what bouncers are for. I tell them that I'll get fired if my boss sees them on me.

The greatest challenge as a bouncer for me is not breaking up fights; it's preventing fights from even happening with intervention communication. The way I see it, if a fight happens, the crew of bouncers aren't doing their job. The pay isn't that good, but I'd rather get the scratch than be out wasting on weed, liquor, and ruined health. The bouncer gig makes me stay fit and alert. Sure I might get hurt, but I haven't up until now. I pray I never do get hurt, and that I'll never have to hurt anyone.

FERNANDO PEREZ

SECOND-HAND HUSTLIN' IN SILICON VALLEY

One man's garbage is another man's treasure—and for
some of us in Silicon Valley, it's an entire economy. I've
been working the garage sale, flea market, Craigslist circuit
for years, and as unemployment checks run out and jobs
continue to run scarce, I've seen a major increase in what I
call "second-hand hustlers."

Families now are organizing yard sales year round, not
just on the weekend anymore to make extra cash. That's
where we come in—to get that stuff, and sell it elsewhere.
You can spot us at your local garage sale, estate sale, swap
meet, or flea market, hustling for those second-hand goods
so we can flip them online.

For over ten years I've been doing the rounds, searching
for rare treasures to bring home, and I've never seen the
competition so fierce. On the hustle trail, I run into more
and more people who are now doing this as their full-time
job. For some people, second-hand hustling is the only job

they have, and for others it has become a second or third source of income.

When a friend was laid off recently, he started buying different electronic items at the local swap meet and yard sales. He's been reselling them online and making a living for his family, sometimes bringing in over $500 a weekend. He says when he works the swap meet through the week, he makes as much as when he held down a nine-to-five job, and he isn't too motivated to go back to the traditional workforce. In this gig he calls his own hours, spends more time with his kids, and has no boss to worry about.

The best second-hand hustlers specialize: antiques, vintage items, electronics, construction tools, basically anything they know that can be sold quickly or is hard to find online. For those who want to start second-hand hustling, always be sure to try and make at least three times what you paid for an item, because every purchase is a gamble. The job comes with a lot of research and traveling. The difference between a pro and a rookie is in the preparation. First-time buyers just go with their gut instincts to determine the value of an item; pros go with their smart phones and knowledge.

The best places to start second-hand hustling are in the more affluent areas like the Evergreen, Los Gatos, Sunnyvale, Willow Glen, or Santa Clara neighborhoods, mainly because they sell better items, and have wider streets with ample parking and bigger yards to browse in. Second-hand hustlers are in such a rush driving around that they don't even stop. They just pull up, browse from their cars, and toss the items in the back.

After stopping at all the yard sales in the nice areas, we usually make our way towards the more populated areas on the other side of town, near the Capitol and Berryessa flea markets. It's not as easy to find parking here, and it seems like everywhere you turn somebody is trying to make some money. Yard sale vendors in these neighborhoods use fence posts and tree branches to hang clothes on, and items are set out on tarps or blankets on front lawns or driveways.

But buyers aren't the only ones looking to cash in. Vendors are catching on. A second-hand hustler's nightmare is stopping at a yard sale where all the items are priced too high. These kinds of bargain hunts are difficult because the vendors know exactly what their wares are worth. Just the other day while browsing around at a big yard sale, the vendor started announcing that all items bought in the next fifteen minutes were 50 percent off. Immediately it created a skirmish among the crowd fighting to snatch up anything in sight.

I haven't made my mega-find yet. Till I do, you can find me, iPhone in hand, at your local sale, haggling with the best of them.

ALEX GUTIERREZ

WORKING AT THE PORN SHOP

Let's talk about sex. That's right, I said it. It's human nature and it's also my job. I work at a porn shop. Adult retail is mostly the sales of adult toys and movies, and the viewing of arcades (which are not really arcades). Now, working at a porno shop, one would think you would be able to talk to beautiful "freak-in-the-sheetz" type women, but from my experience, nope. The action you could get, you really don't want.

Having worked at the porn shop for a year, I can say that the last people you would guess are into porn frequent stores like ours, and that sex sells big in San Jose. There are all kinds of different people at porn shops. I've talked to people from all different backgrounds and places from across the world—New York, Mississippi, Florida. I've even talked to people from Beijing, China.

But there can also be the occasional crackheads, women who are trying to conjure up adult business for themselves, and even everyday drunks bugging customers for change as they walk in and out the door.

But sometimes the customers want more than just merchandise. The offers are crazy. It doesn't have to be a Friday
night for stuff to be crackin'. Any day of the week, some
customers may ask you what are you doing after work,
would you like to go and have a few drinks, or even give
you a personal invitation to their household. None of those
offers are wise to take at any time.

But for some customers, sex isn't the only topic. Oftentimes it's more serious intellectual conversations, ranging from issues in the community or the government to
relationship issues they're having. But most of these conversations all end the same, with them talking about hiding
porn from their wife or girlfriend.

The environment of the shop is kind of like a sitcom.
One time a pastor I knew came into the store trying to
stalk up on some special DVDs. He saw me, got real nervous, and bolted out of the store. And on another Saturday
night, a man walked in with dry, chapped lips that had puss
seeping through the skin. He pulled up a bag and placed it
on the glass frame. In the bag was some hamburger meat—
pounds and pounds of hamburger meat. He then offered
me two pounds of hamburger meat for three hours of time
in the arcade booth. Then there are those guys who come
in the store buying lingerie they say they need for their
"girlfriends" but measure it to their own body and ask if I
could tailor it.

Some customers that come in are actually just trying
to practice safe sex, in that they have an STD, sometimes
AIDS, and don't want to infect others. Actually, one of
our coolest regular customers told me one day he was
HIV positive. I remember thinking he didn't look sick and
always seemed to be in good health. I felt kind of weird

at first, but then I thought, if I was cool with him before, there's no need to treat him differently now. To this day he is still my best customer.

One of the main topics all adult novelty stores should be concerned with is security. Police oftentimes do not respond if we call because of the industry we work in. Last month, I had to fight off a crackhead who was harassing the employees. He tried coming after me, even hit me with a newspaper stand. I retaliated in my own defense and whooped his ass. But still, the point is we had called the police earlier and they had failed to respond. Employees from the shop are not what the police think. Females are not superfreaks and guys are not perverts. We are just people who got bills to pay and families to feed. The porn shop is a good business and means of employment, and working here doesn't make us any different, or put us below anyone else. The porn shop is a job, and the main topic for sales might be sex, but sex sells, and as long as it does, I always have a job.

YAVETH GOMEZ

LIVING IN MY CAR

It's crazy to think that I make more money than most people in the world, but I'm still unable to afford a place to live. I make $58,000 a year and it's not enough to live and be able to eat here in San Jose. That's why I live in my car. Sleeping in my car is my way of fighting high rent costs. I want to save up for my future. I want to be able to have a place to call home—not to borrow and give it back when I run out of money.

I can either go into a lease agreement where I can easily go into a lot of debt, go back to grinding and hoping to get promoted only to be more in debt. Or I can keep grinding but give up finding a decent place to stay due to the fact that getting a place equals being broke as fuck. So right now, I'm enjoying spending a little money, paying debts faster, and am on a somewhat clearer path to saving up for something bigger.

I was born in San Jose, and lived with my family here for more than twenty-five years until we lost our home to foreclosure during the recent market crash. We moved out to Tracy, and while I had a job here in the area at a bank, I commuted back and forth for more than a year.

I got that job as a teller after graduating from UC Berkeley thinking I could move up and grow, but I hated working there. I would go home to my little room that I rented from a Filipino family and play video games with my housemate, a nine-year-old kid who played Grand Theft Auto on PlayStation 3, and I would take out my frustration by killing random people on the street—in the game of course.

After a while, the decision of whether to continue renting the room in a shitty rundown area of Santa Clara versus commuting from Tracy was made for me. I got fired from the bank, moved back to Tracy, and started looking for a job that I somewhat liked.

I knew I wanted to help people and I started getting involved in community nonprofits because that was always something I enjoyed. Eventually I applied for a job at a union, where I am currently working now.

The only sad thing for me is that others don't know I've been living in my car. I started to be more open about it, but what I hated about telling people is that they assume this is not my choice and they try and help me only to realize there's really nothing they can do.

I would go home to Tracy to visit my parents and go back to my room I have there, thinking to myself: damn, I could just be sleeping in my comfortable bed in Tracy if only I had the discipline and will power to do the four hours of commuting every day. My parents didn't even know I lived on the street. If they found out they would probably have a shit storm.

Keeping all this secret took a mental toll. I constantly lied about my situation. To the question, Where do you live? I would tell people I commute, or I'd tell people I rent a room—I would tell them anything except that I live in my car. My friends knew I lived in my car and my close friends would be shocked, like, wow, how do you do it? They'd be worried, but otherwise they'd admire my tenacity.

If the government doesn't have the will to help out the depraved in this city, choosing to chase the homeless out from one spot to the next without real long-term housing solutions, they can at the very least have car-parking sanctuaries and make living in your car legal. It can be very simple, like a neighborhood park where people are allowed to park their cars overnight and feel safe. Sometimes instead of just fighting the inequities of the Valley, you just got to come to terms with them, and make it work for you.

Tips for Sleeping in Your Car:

1) Park in a safe, chill neighborhood. Make sure you don't see any signs saying that you're not allowed to park, and if you see a neighborhood watch program sign, just be aware that you might get told on if you are spotted. No one will notice you if you don't make it too obvious. It would be cool if you had tinted windows. I know some people may live in their RVs and park outside of Home Depots or places that are mostly industrial and don't have security watch.

2) I parked around a park where I could use the restroom in the morning, but I wouldn't recommend it since

those restrooms are disgusting. It also helps to have baby wipes as a shower substitute, but that is not exactly an effective alternative. Have a sleeping bag and/or a thick blanket with a couple pillows to sleep comfortably. Keep your clothes on even when you sleep because you never know if someone will come and wake you up to kick you out of your parking spot.

3) I know it's illegal to sleep in your car in most places. Don't let your battery run in the car unless you have jumper cables and you don't mind wasting gasoline or energy. Bring a nice book if you're parked under a street lamp. Don't make it obvious by parking and putting all kinds of covers on your windshields because that just looks tacky and will get attention.

4) A constant challenge is having nowhere to cook food. If you're going to be extreme, have an adapter that can connect to a Foreman Grill or some low-vault pressure cooker, or steamer to cook all your stuff in. Park near a Safeway to stroll in and buy what you're going to cook and eat right away. Also, a Safeway is open twenty-four hours, so you can go use the restroom if you have to.

5) I didn't have a gym membership so I would sneak into a local college and take a shower in the locker room whenever I had the chance and I was near campus. I would have myself a free breakfast at the buffets where they had a free event or forum, or I'd pretend I was a guest at a hotel where they give a complimentary breakfast. I know that's wrong, but that was sometimes my little scheme to save money.

LAMAR WILLIAMS

YOUNG, BLACK, AND HOMELESS IN SAN JOSE

My birthday is coming in a couple of weeks. It'll be my twenty-fifth, and it falls on a Saturday. Saturdays have always been kind of important to me because ever since I was little, they were my day for cartoons and video games. It's how I'd like to celebrate—a bunch of cartoons, some people around, drawing, a pot of chili, and maybe a few beers.

But first, I have to find a place to stay. Currently, I sleep on the floor of some friends' apartment, and I have no key. This is a blessing, and I've had worse accommodations (try the back seat of an old Toyota 4Runner with no back window in the dead of winter). But there is something maddening about not being able to come and go as I please. In any event, I've only got this arrangement until March 1, at which point I'm back to my own devices.

Now, before I landed in my current stitch, I checked out every state program I could find. Help for immigrant refugees, help for families, for women, for long-term homeless, for minors—nothing for an able-bodied man down on his luck. And when I say "down on his luck," I mean it. I've been unemployed, with a few short stints of employment, for three and a half years—so long I don't qualify for unemployment.

And I do everything. I write, I edit video, I install car stereos, cook, clean, drive, shoot pictures and video, garden, watch dogs, teach, program in C++ and Python—but no job has ever worked out for long. I've got a smell on me or something—too much ambition, too much of an independent spirit. Or maybe it's just a smell?

I've been laid off from three jobs across three industries—electronic assembly, telephone connection, and video editing—over the course of one year. The assembly job lasted two and a half weeks, at which point a manager told me the temp agency that hired me said they were letting me go. To this day I don't know why. The telephone work dried up as a result of a misprint in my boss's phonebook ad. She just stopped getting calls during what should've been the busiest time of the year, and we didn't figure out what happened until I was about a month into work as an editor. The editing work went away when my studio's contract providing me with steady work fell through.

This time last year, when the California unemployment rate reached 10.5 percent, the rate for black males was 16.3 percent. My resume was everywhere, and I can't remember any actual interviews from the period. I moved into my grandmother's house, and helped with errands and maintenance while I waited and prayed to find myself back on my feet. One in five black males share my story, and it wasn't too long before my uncle, a truck driver with more years of experience than I've had birthdays, joined me as a statistic staying under my grandmother's roof in Poway, a small town in north San Diego County.

My father, who moved cross-country not even a year before because of a promotion, had at this point changed his tone in our conversations, from "Son, you need an income, you have to go find something, anything," to, "Times are hard, and I feel for you." He'd also gone from telling me when I was in high school that he didn't want to hear about me enlisting in the military (*Boyz n the Hood* was one of the first movies I remember seeing in a theater, and the line "the black man has no business in the white man's army" had always been a favorite of my dad's), to being a bit more accepting of the navy, since at least I wouldn't end up in Iraq.

Meanwhile, I'd been filling out applications, sending out copies of my resume, writing cover letters, and putting hours of research into companies I'd never hear back from. And the statewide unemployment rate for black males climbed to over 17 percent, while the rate for white males was at 9.4 percent. Shit was rough.

It was around this time, sitting in a community arts center, which has become a respite for other creative twenty-somethings who can't find stable employment, that I wrote on a sheet of paper, "If no one else will hire me, I might as well hire myself." Things began to turn around. I set to work on projects that were important to me, and it resonated with the people who were already there. I'm more creative than I've been in a long time, and feel better about what I do every day than I ever have.

But it's a little too late to turn the bus around. I won't have any help from the state, and I'm fresh out of floors to sleep on. I should be more afraid, but it just doesn't bother me so much—I know something good is going to happen.

On any given day, I probably don't have a dime in my pocket, and with freelance gigs, it sometimes takes a while for checks to be turned around. Plus, my credit is just bad enough that any place with a rental application probably isn't an option.

But I love it in San Jose, and I've got work to do here, no matter what my situation. The people have been kind enough to take me in when I had no place to go, to feed me when I had no food to eat, and have made for good company when it came down to talking shit over a game of cards, so I owe them every scrap of energy I can muster. Thank you, San Jose, for being my home.

ALEX GUTIERREZ

MOTEL 22

It's just past two in the morning and the wind blows fast and sharp. Everyone waits at the Eastridge Transfer Center for the bus 22 headed for Palo Alto. We all gather around as it pulls up near the edge. As the door flaps open, everyone pushes their way in like salmon swimming upstream. The bus driver starts to yell, "Back up, either line up straight or no one gets on the bus!"

We all back up, hoping to get out of the cold as soon as possible, flashing our day passes. A few pay in cash, trying to negotiate their way for a few free rides, at least until the morning.

The 22 bus travels 24 hours a day, 7 days a week, 365 days a year in and out of the streets of San Jose, Sunnyvale, and Palo Alto. The streets are lonely and quiet throughout the middle of the night, and the bus ride can last at least two hours between San Jose and Palo Alto. A lot of people call the bus line "Motel 22," because if you don't have a place to sleep that's the only place to go.

Two people get turned away for not having the fare to ride, and the rest of us set down our belongings and get in position to get some rest, at least until we have to switch buses back to Eastridge. It amounts to two hours of sleep back and forth per trip. A whole night usually comes out to about five trips a night, switching buses every so often. I sit in the middle, on the left side of the bus, snatching whatever sleep I can before it's time to switch at the last stop.

I wake up a few hours later with my cheek pressed against the cold glass of the window. I look around the bus to see who is around me. I notice the woman talking to herself sitting behind me, the group of kids in the back of the bus talking about where they're "gonna find some bomb at three in the morning," along with the guy trying to do magic tricks, bothering whoever glances at him.

The Valley Transportation Authority (VTA) security gets on the bus looking for whoever is asleep to tell them at least to sit up and make room for whoever is also getting on the bus. While going through the passengers, the guard smacks one of them too hard, and an argument breaks out. The conflict doesn't last very long because the guard threatens him, saying that if he keeps talking, he won't be riding on any VTA bus at all, and will be stranded in Palo Alto the whole night. The passenger stops arguing.

During a bus switch heading back to San Jose, three others and I step off of the bus and started crossing the street. As we cross the street, a VTA security officer pulls up in front of us and asks us why we crossed the street on a red light. In my mind, I'm thinking it's not about the red light at all, but more discrimination against homeless people who ride Motel 22. He threatened us with not being able to get on the bus back home to San Jose. The matter was squashed and we headed back to San Jose at the Eastridge Transit Center, then we waited until it was time to catch the next bus back to Palo Alto. It can be a tiring practice, but at least the bus is warm and shelters us from the cold.

JUSTIN COLLINS

THE TAO OF HUSTLE

The term hustle *conjures thoughts of quick money, which is all too often a necessity in Silicon Valley. But for some, hustle is a philosophy, a way of life, a Tao.*

The Tao of Hustle is...
Knowing true strength is knowledge.
Remembering those who fell and why.
Asking not what you can do for your hood
But what your hood can do for you.
Not talking over the phone; someone is always listening.
Putting family first, once they're gone you'll miss them.
Knowing when saying less translates to more.
Staying loyal when only you will know.
Seeing without looking.
Knowing when to walk, quickly.
Not incriminating yourself through boasting to a friend.
Never doing anything you're not willing to answer to.
Not promising things you can't do.
Showing more than you were shown.
Listening to an O.G., and not interrupting.

Getting back in school.
Learning your lesson.
Not being greedy.
Having mercy.
Being the first.
Remaining honorable.
Having nothing to hide, or working towards it.
Feeling good as you think, "I could have just came up, but
    my life's too valuable to find out."
Having verifiable earnings to show the tax man.
Not cheating on the "once in a lifetime girl."

SHANA WHITE

DON'T CALL ME LUNCH LADY

I hate the name "lunch lady." It makes me think of a fifty-year-old, plump, white lady wearing knee-high socks and an apron. Um no, that's not me.

I am a chef at a private school in East Palo Alto called Eastside College Preparatory School. We have 340 Students from sixth through twelfth grade. Looking at the filled cafeteria every day makes me wish Eastside was around when I was in school. Back when I was growing up, East Palo Alto didn't have a high school.

A lot of people don't think about how important it is to be the "lunch lady," but it takes as a tremendous amount of discipline, and the patience of a monk.

When I first started at Eastside seven years ago, I began as dishwasher. Before I was hired here, I had been unemployed for a few months. In the Valley, that can cause a lot of anxiety. My soon-to-run-out unemployment check barely covered rent and my savings was getting pretty low. So to help out with things like food and gas, I went to work with my mom doing house cleaning and even did yard sales for a bit. That's when a friend of my cousin told me that her job was looking to hire a dishwasher at a private school, where she works. So as soon as they gave me a chance I signed the application, and have been there ever since. It wasn't easy in the beginning. While I've been there I've learned a lot and seen a lot. But through all of it I enjoy what I do there. The students, although rowdy, are also very respectful kids.

Of course, we do have students who like to test the boundaries. It's usually the middle-school kids who decide to take the fire extinguisher and spray everything, or like to stand up on the sinks and see what happens when they press the red buttons.

We have a great and diverse crew in the cafeteria. We have: Ritika (chef assistant/line cook) from Fiji—a young twenty-eight-year-old mother of two. Cole (evening chef), a twenty-eight-year-old who came to California from Oregon to find something new; he enjoys cooking, and has participated in the Iron Man Triathlon. Edgar (dishwasher), a twenty-five-year old who came from Nicaragua also to start a new life and work. Edgar is also a DJ who spins everything from reggae and pop to old-school R&B. And while it is a solid crew, as the boss, it comes down to my responsibility that the kids get fed. I remember one day where for various reasons all of my staff weren't able to make it to school. So that day I had to make a full salad bar, pan and bake three hundred–plus pizzas, break down three pallets of food shipments, serve, clean up, and wash dishes.

Each day we feed 340 students plus staff. I plan the lunch menu each month, cook, clean, and serve. I do it all. I cook everything from pizza, lasagna, and roasted chicken to plain-old grilled cheese with tomato soup. We have about twelve people who are truly vegetarians, three students who cannot eat dairy products, some students who don't eat poultry, and most all of the students who do not like vegetables.

The students work really hard; they start at 8:00 a.m. and go all the way to 5:00 p.m. Back when I was in school I thought 3:30 p.m. was late. That's why I sometimes give them a treat, like ice cream floats once in a while, or have nacho day on some Friday.

But lunchtime can be like a showdown between the cafeteria staff and the students. They are rowdy, picky, opinionated—but we can give it as much as we take it. We get a lot of students who know us as much as they do their teachers, and they ask us about our days and we do the same.

We do get the picky eaters who don't like what's on the menu, or act like they are at a restaurant. But it's expected. Actually, my real stress moments don't come from the students, they come from other adults who visit the school: the health department people and the food-delivery companies.

Usually, before the health department comes into the cafeteria I'd get a warning call from the secretary up front, but now these days they've gotten slick and come in through the back door. We always pass inspection, but it's just that when they pop up, it gets scary. And as for food-delivery companies, they look at us like the big score, and are always trying to win us over from their competition. It's the politics of the job.

Now that people know I run the kitchen, I always have a family member or a friend ask me to get them in to work. They are always telling me how hard of a worker they are, or were. But the thing I tell them is, it's not the job that you should worry about, it's me. But I couldn't do this alone; I run this ship with the help of every teammate. People say I have to take it down a notch, but if we aren't precise, prepared, and pushing as hard as we can, these kids don't eat.

CESAR FLORES

SERVING UP SILICON VALLEY

I am a server at a local brew house. I've served in different parts of the state, but serving in the Valley is wilder—staff party harder here.

While we are at the restaurant, the vibe isn't much different then from at the parties after work. It seems like everyone is on something to calm their nerves. Some people take a couple of shots of whisky before work, a lot of people smoke weed during breaks—most customers would be surprised. Chances are, whoever is serving you your plate of pasta or hot wings is high. It's just how we get through the madness of the shift.

There was one time when I was at work and there was some kid that was really helpful. I thought he was just being super productive. I asked him if he got a really good start to the day. He said he actually had a huge hangover because he went out the night before. I chuckled. He was on his phone a lot throughout the shift, and I asked him if he was talking to a girl or something. He told me that he was actually calling his friend to bring him more coke to wake him up. He ended up getting fired a couple weeks later, but that's just because he was slipping way too much.

There are usually only two types of servers—hustlers and slackers. You pick up really quickly on who does and does not do their job correctly. It's funny to see the hustlers from the slackers; you can automatically tell. The hustlers come in looking like they're ready for war or something, stealing tables from other servers. We call them table sharks. The slackers and newbies usually take back seat to the hustlers for the simple reason that they don't care or are oblivious to the fact that the tables taken were theirs in the first place. Some just get punked out of tables, thus money, that should be theirs.

There's an adrenaline rush that comes in when one is serving that I have never felt before in my life. For people who have never served before, let me break it down like this.

Imagine having one of your friends asking you for a favor, and as you say yes to your first friend, you get a call on the other line. Your other friend asks you for another small favor. You feel like you could handle all of it, no problem. But then a third friend calls and asks for a quick favor. You feel like you'll be stretched out a little thin, but why not. The fourth friend calls, you pause and start to see how many things you have to do. Imagine not being able to say no to the fourth friend, and it keeps accumulating, until ten friends are asking for favors. That is what it feels like with your tables, except these people are not your friends, and they let you know that very clearly.

But what is unique is you are often witness to very intimate moments in peoples lives. I've seen more break-ups than I would have liked, guys walking out making the girl pay, girls walking out and throwing their cocktail on their significant other. Relationships, flings, affairs, built and destroyed in front of me.

I remember one time when the table was so tense, I ended up being pulled in as an ally.

It was a big family who were not talking to each other, and when they did they were yelling. It seemed like one of those court-mandated family visits. I broke the ice by making jokes and joining their forced family dynamics just as a favor, because the husband looked like he could use some help. I ended up praying with them after they asked, and to my surprise got a two-hundred-dollar tip on a sixty-dollar tab!

But serving is, without a doubt, hard work. At first it drained me to the point where I had to sleep twelve hours in order to recuperate from the day before. But after getting used to it, I don't think I would feel comfortable not having that stress in my life—the chaos can get almost meditative. It's so hectic that it makes you think of only that moment, all your problems, fears, hopes disappear in this vast energy, and for the next four or so hours you are doing what most Zen masters take their whole lives to teach, the moment. I just have a tray of mozzarella sticks when I am in it.

This same reason is why being a server is such a curse at the same time. Being constantly in the moment makes you forget about the rest of the world. Waiting will drain you. It seems like in the last two years of serving, I have just a handful of actual memories. You don't really realize how much time you spend at the restaurant because you're hustling to make money so hard.

The restaurant itself is a whole different story; it's actually kinda gross how things get taken care of. Plates are rarely clean enough to give straight to the customer; I'm always wiping gunk off of "clean" plates. People sneeze, I don't see them wash their hands; people put dirty plates away, don't wash their hands, grab straws with the same unwashed hands. I remember one day when I first started, someone was running food and accidentally dropped a hamburger on the floor. He immediately ran it to the back, wiped it off, and took it to the poor lady. The five-second rule reigns in restaurants too.

I've never seen the cooks spit in anyone's food. But I have my suspicions, and the targets are not always just the customers.

One cook didn't like me. I never knew why. I later found out it was because I was picking up "dead" food. Dead food meaning food that went cold because someone didn't take it out to their table. I had to stop ordering food from that section of the kitchen because every time I ordered food and went out to pick it up, the cook gave me an evil smirk.

I don't know if he is putting any special sauce in the food or not, but that's just part of being a server. You have to keep your feet moving and your eyes open.

ADRIAN AVILA

AMERICA, MOM, AND BBQ

My mother and I came to this country in March of 1990. I was five years old when we left Mexico City, Mexico, and headed up to the Golden State of California.

We settled in San José, the heart of the Silicon Valley.

At the time of our arrival, the city had a population under 800,000 and was making a transition from a historically agricultural town to a high-tech city. With a limited skill set and a language barrier, although my mother was intelligent enough in Spanish, she couldn't enter that tech economy unless she was cooking food for the high-tech job market. Metaphorically, she had to enter the industry through the back door that led to a kitchen to cater to high-tech industry workers. But technically, that kitchen wasn't in the building, it was parked outside. Her first job in the United States was working on a lunch truck, going from job site to job site feeding those who were building the technological revolution.

In Mexico City my mother had been a secretary to a US-based company for over ten years. Moving down from officer work to working a lunch truck was a big change for her, but she knew that you have to start somewhere.

That food truck job was a start to a long career in the food industry for my mother. Her next job was packing fruit at a factory in west San José. The hours were strenuous and the work was all hands on. One of the perks of the job, though, was buckets of fresh-cut fruit, which I was particularly fond of. My mother found comfort in being able to provide the fresh fruit to me. Being an only child in a little family of two, I watched my mother work hard to put food on the table. So fresh fruit was a big deal to me, especially on a strict budget.

With each job brought a new change. The joy of the fresh and healthy fruit was replaced a year later with the sweet smoked flavor of good old American barbecue. My mother landed a job at a barbecue restaurant starting off at minimum wage, four dollars and twenty-five cents an hour. Like most restaurant jobs, it was tough. The days can be long and the body can grow exhausted from the repetition of the day. Imagine, a five-foot, one-hundred-pound Mexican woman lugging around big beefy plates of hot, juicy full racks of ribs, washing pots and pans, and cleaning up tables until the day turned to night.

This was a routine that I witnessed firsthand when I began to work alongside her around the age of sixteen. I remember one day starting work around four o'clock in the afternoon. My mom had already been working since 11:00 a.m. and didn't get off until 10:00 to 10:30 p.m. It being a Friday night, we knew that the place was going to be packed, and it was. Person after person, plate after heavy plate, my mother was on full beast mode. At one point in the night she moved me out of the way and said to let her take care of the orders, and I should just go to the back to help wash dishes. She was letting me know I wasn't fast enough to work the front line. I didn't last there for too long; the place just got to me. I knew it was work that was just not for me. The time I spent working alongside her, though, made me appreciate her work ethic even more than I already had.

No matter how tough the day was, she managed to genuinely find a way to smile—something that customers greatly appreciate about her. After twenty-three years on the same job, she is still smiling. Not only has she been able to live on her own, debt-free in one of the most expensive counties in the state, but it turned out that working at this BBQ restaurant would have a huge part in saving her life.

My mother was diagnosed with ovarian cancer in early 2011. It all happened so fast, I didn't really have time to react to the diagnosis. But deep inside I knew that things would be okay, I knew she was covered and was in God's hands.

Millions of Americans don't have health insurance, and my mother was lucky enough to be covered by her work. She has had health coverage for over fifteen years, thanks to a job that most people would not associate with good health insurance. Taking into consideration that my mother is still undocumented, I think it a miracle that she had that type of protection in her life at a time when millions of Americans couldn't say the same. When it came time to deal with the cancer, she was well equipped to do so and beat it within a month of being diagnosed. An act that my mother understands to be the work of good karma and God.

She was back at work lugging around full racks of ribs exactly one month from her diagnosis.

At times I sit back and think about how hard it must be for my mother to keep smiling after so many years of the same hard work. I think about how far my mother has come since we arrived here in the US and how lucky my mother was to land a job that provided health insurance that literally saved her life.

My mother continues to work hard, not complaining and making the best out of her work situation, which at times can get hard. But if it wasn't for that hard work, she might not be here today. She has truly become an American in the sense that she came to this country with nothing but me by her side. She is still not welcomed by this country, although she continues to feed it and work hard for its people. Being undocumented in this country is hard work, but as she always told me, nothing comes from easy work.

CECILIA CHAVEZ

TO THOSE WHO MAY HIRE MY DAD IN THE PARKING LOT

I want the person who sees my father to know something about the man you're about to pick up to help you with your weekend home improvement project. When you turn into the parking lot of the hardware store where he awaits for a job, don't get discouraged by his age or appearance. Don't deceive him or trick him into doing your job without you having the intention to pay him, like many others have done. He is a hardworking man who has endured many long years of strenuous jobs and is a seasoned veteran of construction. He knows what his job is worth, but because of his years and status, society has pushed him aside and forces him to seek work by standing in the parking lot of this Orchard Supply Hardware store.

He gets up at five in the morning every day, seven days a week, and stands in the parking lot of OSH to see if someone needs extra help in any home project that they are working at. He wakes up hoping that someone will offer him a job that pays enough to help out my mom with the rent and bills. Through his rejection of his age, he is still willing to endure the humiliation and continued search for a job by standing in front of hardware stores.

So let me tell you who you are looking at through the window of your car.

I am the daughter of a "jornalero," a day laborer. One of the guys you see that stand around Home Depot, Orchard Supply Hardware, or any hardware store that people may come to needing help to build something. My father is Celestino Chavez; he is a longtime immigrant from Michoacán, Mexico. He first immigrated to the US in the early 1970s, when crossing the border wasn't as hard as it is today. He was a migrant worker, traveling back and forth to Mexico to his family when work was finished in the fields. Once immigration laws restricted his flow, he decided to settle in the US because he saw that there were more opportunities here. This migration took place twenty-two years ago. He is currently sixty-four years old and still has the same drive to work as he did when he was younger.

My father has always had a strong work ethic. He started working in the fields before he finished the sixth grade in Mexico. When he was eighteen years old, his dad was killed by the military and he had to step up to plate and become the head breadwinner for his three brothers and mother. Once he got married he continued to support his mother and new family. But life in Mexico was not very easy, and jobs were becoming scarce. When he decided to settle in the US, he brought over his entire family so that he could provide a better future for all of us, six kids and my mother.

When our family settled in the US, my father was working as a handyman and taking care of the horses in a private ranch in the Alum Rock Hills. He was always working
extra hours but never receiving the extra pay. He was aware
of it, but because of the fear of losing his job, he would
not complain. After dealing with his unscrupulous boss for
more than seven years, he finally decided to leave that job
and apply to other companies. He was able to hold a job in
a respectable construction company for a couple of years
because he was using his brother's identity. This construction job paid him well and our family was able to purchase
a home. Everything was going good, too good to believe.

They let him go once they found out about my father's immigration status, and as heartbreaking as it was
for him, our entire family's world crumbled. He was the
head breadwinner and the rock in the family. Everything
and everyone depended on him and his job. Once he lost
that job, the bad news had a domino effect on all our lives.
We were unable to continue paying the mortgage on the
house, and the car payments were always late to the point
that his only car got repossessed. Besides the material loss,
the hardest thing for my father was feeling like he could
not provide for his family.

It took years for him to get his confidence back and
get back on his feet fully. He would search for jobs even if
he had this sense of depression reigning over him. Losing
his job meant more than losing a paycheck. For him a job
meant being able to provide for family, being an independent individual, feeling resourceful, and feeling useful.
His self-esteem and morale took a dive when he was left
unemployed, but he was able to lift himself.

My father chose to continue being on the workforce by becoming a day laborer. He decided that he would continue working no matter what. If other companies did not want to hire him, he was going to try, even if it meant waiting in a parking lot all day.

GABRIEL

AMERICA FROM THE BACK OF A PICKUP TRUCK

The morning is still somewhat dark and the phone rings about 7:00 a.m. I answer it and he says that he's on his way. I rush out the door to go to the liquor store before he comes. Across the street, to the parking lot in front of the liquor store. I see a sign that says "No Loitering," but it's almost covered by all the Mexicans waiting to get picked up so they can go to work.

I wait another few minutes and not too much longer. I hear the truck. In the truck I feel the warmth of the heater and the calmness of the radio. I sit in silence after closing the door. The Boss heads to the highway to pick up Enrique who is probably waiting for us. I see him as we pull up to the curb and he greets us, getting into the truck. The radio station is always the same: religious talk from the DJ along with songs from Vicente Fernandez and Pedro Infante, and the requesting of songs from people that call the station. The things that always cut through to us are the news bulletins, and not just the reports about the traffic. The thing about the news is that it always has something about immigration, workers, and the economy. Callers talk about the morality or legalities of immigrants and all those reasons why they deem the immigrant a danger to the nation. Enrique makes a comment about the hypocrisy of letting an immigrant make salads for rich people and at the same time calling him a thief.

As we continue north on Highway 101, we are passed by rushing cars. I wonder about the current political events that affect all of us in the in the truck, except for the Boss. We are all construction workers, undocumented, and most of the men are older than me. We all wonder what will happen to the fate of all of us, whether we will always struggle through even harder times.

While getting out of the truck, I look at all the hills with signs that say "Jesus Saves." I start thinking of the virtues taught to me as a child and the prayers we all learned before coming to this country. In my head are the few verses that the church tried to make us memorize, the things I cannot say anymore. These prayers are what come to mind while working in the sun with the scratches and calloused wounds on my arms.

We work the days in the silence. These men's thoughts float in the heat of the day. They are of English words like "minute men," "real ID," "licenses," and "terrorists"—words from the morning radio. We are all pouring down sweat and the bitter burning taste flows with the words that we have all become familiar with—"illegal alien."

These mansions that we work in throughout the highways are like pyramids that are erected in grand, majestic reflection of society. And we are the mason builders that have gone from city to city, town to town, leaving only the trace that it is actually our work. On TV we always hear, "Aren't they stealing from our country?" The question is ironic as we open our checks and look at the social security money taken away from us that will never be returned, leaving the owner of the number reaping the benefits. Is this the way crooks steal from the nation, by having the government take their hard-earned money? Nonetheless, it is the price we have to pay.

From the highway, the rows of cars pass during the afternoons and in the mornings. Those drivers all have licenses that we are told we cannot receive. Yes, a license to drive is a privilege that shouldn't be taken for granted and people should be properly trained before they get on the road, but simply taking away the right to drive when people have families is absurd. It reminds me of the time when people weren't allowed to vote because they didn't know how to read, and if someone gave them the opportunity, these people could be honorable citizens. Plus, could a terrorist attack really happen because these construction workers were given licenses? Does America know that we are like all people, that we work, eat, rest, and know virtue too?

We are slowly building another American town— companies and houses are moving to the land we prepare. We work hours into the night, and going home we stop at the gas station where we buy something to eat and drink and head onto the road. Watching the night slowly travel with my body resting, I listen to the radio and stare at the hills. I look for the sign that says "Jesus Saves," but it is too dark.

ANGEL LUNA

AMERICA IS IN ME AND SO IS MY ACCENT

I was two years old when I left East San Jose to live in
Oaxaca, Mexico. I returned to the United States when I
was fifteen, and was ready to leave my Mexican life behind
and start my American one fresh. But today, I want to keep
my accent.

Now, to hear me, imagine your favorite rapper, except
with a Mexican accent. The letter *g* gives me some trouble. I will try to say "What up, gangsta?" and it sounds like
"What up, ansta?" I pronounce *v* like *b,* which becomes an
issue because I'm a "bideo maker." Having Wienerschnitzel
as your favorite restaurant doesn't help. People still understand me, though, and I use a lot of body language, too.
Sometimes it even helps, like this one time I called a girl
"chubby" but she thought I said "shorty" and was pleased
with the compliment.

When I left Mexico, it was the hardest day in my life.
I left my home, my friends and all the things that I knew.

Even at the border, I got hit with a reality check. At the checkpoint, immigration officers told me they wanted to question me. I showed them my papers (all legal, by the way), but then they asked me a question in English. At that time I knew no English. I was pulled into a detention center and was interrogated for hours. It felt like I'd stepped into hell. I saw old ladies getting hit and heard officers cursing like crazy. I didn't know what "beaner" and "wetback" meant at the time, but by the way that they were yelling it, I knew it wasn't good. It was the start of a feeling that the America I left years ago didn't want me anymore.

Once I settled back in San Jose, I tried to learn English as fast as I could. I didn't want anyone to have an excuse to discriminate against me or look at me funny. I was a sophomore in Overfelt High School in East Side San Jose for my first day of American school. I had to do a test to see what class I would be placed in. I scored 100 percent in the Spanish part and 10 percent in the English part. The teacher was impressed, so she gave me the choice to go to a higher language arts class or stay in the beginners one. I took the higher class, because I had a hunger for learning English.

My ESL class was full of "paisas" (somebody from your same nationality), and I got updated on the class by a classmate. I asked him how long he had been in the States and he said six years. I was shocked—six years in this kind of a class and it sounded like he knew just enough English to get a meal at Mickey D's. My goal was to be out of those classes in less than a year and become a "regular person." During that year I did hella good at school and, soon, almost all my classes were in English. All my folks at school and at home were telling me how surprised they were at how fast I was integrating into America.

One of my cousins told me, "Damn, man, in few years you ain't even gonna have an accent anymore. You gonna sound American." The comment made me proud to some extent, but underneath I was starting to be ashamed of myself. A little later, I even made a comment about the speed of my transformation. I told a friend that in about a year I wouldn't even have an accent. She told me, "You're tripping, having an accent is part of you!"

I was stunned by her response. She was right. Trying to be American also meant my slow cultural self-destruction, and my accent was the last piece standing. I didn't realize how ashamed I had become of my roots. I had lived so many days trying to not to be myself, running away from names like "beaner" and "wetback" that I now felt protective of my accent. Having an accent is a beautiful thing; it's what makes me unique and announces what culture I am from.

Now, I proudly keep my accent. But even though the way I sound is cool with me—plus, girls think it's sexy—it's not all gravy. It is very hard to explain to employers that having an accent does not mean you are an idiot. The good thing is, since I want to work in technology in Silicon Valley, a lot of my potential employers may have accents themselves. The truth is, everybody has some sort of an accent, though not all of us know it.

DAVID MADRID

TRANSLATING AT TARGET

I just landed a job at an East Side Target store here in San
Jose. Out of the roughly thirty workers in my department,
everyone, except for maybe six, are Latino. Out of all those
Latino workers, the majority are Spanish-only speakers.
Only two don't speak any Spanish at all. I'm one of them.
Not speaking Spanish has been a problem from day one. I
have always felt the social segregation between Chicanos
and Mexican immigrants at school and on the street. But
now, in this Target warehouse, that feeling is undeniable. At
times I feel unwelcome, and even animosity towards me for
no other reason that I can see except for me being Chicano and I can't speak Spanish.

Throughout the different jobs I've worked, I have often
seen workers discriminated against because of language, but
usually it is because they don't speak English well. Here, I
feel discriminated against by my peers because I only speak
English. It's bad enough that I feel uncomfortable around
the usually white bosses who look down upon me and
treat me like a foreigner, even though my family has been
here for generations.

All over the country, there are Chicanos like myself
who don't speak Spanish. And language alone can lead to
misperceptions, segregation, and conflict among our own
people, even in Target stockrooms. My Target situation is
soon to be a California situation. Four out of ten children
in California now have at least one parent born in another country, the highest percentage in the nation. The jobs
many immigrants are getting are also the ones Chicanos
are already in: megastores like Target and Walmart. At work,
sometimes I feel uncomfortable and have a strange feeling
of not belonging, from not understanding the conversation
and laughter around me during work, to being ignored at
lunch and breaks. On my first day, a couple of lady coworkers attempted to have conversations with me in Spanish as we went about our work, and to their surprise found
out I couldn't speak the language. That would be the first
and last time that I would ever have any contact with them.
They wouldn't even look my way as I passed them down
an aisle. I would even have trouble understanding some
of the morning supervisors who would use Spanish when
giving instructions to our department over the loudspeaker.
Sometimes they come walking up to me giving directions
in Spanish before they catch themselves and finish telling
me what they're saying in broken English.

Growing up, my Chicano family and friends who only
speak English usually had similar encounters with Spanish
speakers. Usually at work or out shopping, a Latino would
come walking up speaking Spanish assuming that I understood, and by the puzzled look on my face, they would realize that I might not understand. If they didn't simply walk
away, they would ask, "You no speak Español?" When I'd
answer "No," they would usually do one of two things—

either smile in amazement or look at me with disgust. Both of these reactions might result with further questions like "Por qué?" or "Aren't you Mexican?" This situation might even end up with an insulting "pinche pocho" as the person walks away.

If I did speak Spanish, the first thing I would do is break down the idea that if you are a Mexican and you don't speak Spanish, you are ashamed of your culture. That misconception underlies the tension. That's why they consider me a "pocho"—whitewashed. And for most of us, this is far from the truth. I know many Chicanos that feel ashamed for not being able to speak the language and who are making efforts to learn, not to mention studying and embracing the culture and history of our people. I can't deny the fact that there are some Hispanic folks out there who are embarrassed and may even look down upon their own "raza," but that doesn't mean if we can't speak Spanish that we all feel that way. Why don't I speak Spanish? You know, I even asked some other Chicanos who are in my shoes how come they don't know Spanish, and the most common answer I heard was, "My parents didn't teach me."

That answer hit home, because it's my response too. My parents don't speak Spanish because my grandparents did not want them to speak it, not because of shame but because of fear. My grandma told me when she was a little girl growing up in the Central Valley, she would get punished by teachers for speaking Spanish at school. You better believe there were no ESL (English as a Second Language) classes for our grandparents and parents growing up. Throughout California and the Southwest during the thirties and forties, it was common to see signs in front of restaurants and stores that read, "No Dogs or Mexicans allowed." My grandpa told me that back in those days, you could even get kicked out of some places for speaking Spanish in public. A lot of our families have endured generations of racism here in the United States and differences of language have made it easy to fall victim to prejudice.

So it must be understood. For a lot of us Chicanos, not speaking Spanish is the result of our families being the victims of a racist American society that dates back to our grandparents and years before them.

At work, I wish those two ladies that first approached me, and all the other Spanish-only Latinos I have run across who look down upon my English-only language skills, knew all of those feelings and history behind my very simple answer—"No hablo Español." I wish they knew that in many ways, I can sort of feel how they do when it seems like everyone around you is speaking in a foreign tongue. But there is just one thing we must keep in mind. Even though we may find differences between us because of our imposed language barriers, we still need to remember what is similar about us. That is, regardless of what language we use, our big white bosses all look at us the same.

VICTOR SALDANA

ONCE A TEMP, ALWAYS A TEMP

A week before graduating high school, me and my friend decided to go to Hewlett Packard and fill out applications for employment. One of my career teachers had said HP was a good place because you could advance quickly.

When we arrived at the security post at HP we asked for human resources. Instead the guard handed us a paper. We were, like, "What's this?" It was directions to a temp agency called Manpower in another part of town. HP said we had to go there if we wanted to work for them. I didn't know it then, but I was in for a work experience I never expected.

It turns out a lot of people working at entry level in places like HP are temps. In this one big company I used to work for, all the permanent employees wear blue badges. All the temps wear red. At break time, when everyone goes outside to eat or smoke or relax, all you see is a sea of red badges with just a few blue ones peeking through.

Most of the temps who did heavy labor at the Hewlett Packard warehouse were young men. Our job title was Box Loader or Material Handler. Women at the warehouse did jobs like Accessory Load and Assembly, which means they didn't have to lift as much as the men. But they had to have fast-moving hands.

You had to stand up straight all day in front of a conveyer belt. The printers came to your left and you had to pick up the printer and place it inside a box to your right on another belt. After you put the printer in its box, you pushed it down Accessory Load, where they would put the manuals and cords inside the box.

We never got bored because we always had conversations. But our hands and backs would hurt after a while. One time, we asked for gloves and they gave us 100 percent cotton gloves, which meant we couldn't put little serial number stickers on the printers because the stickers would stick to our gloves. We asked for thinner gloves but Manpower supervisors told us those gloves were to protect the hardware. I got a hold of a pair of thinner gloves and I'd switch off with some of my friends. But then my supervisor told me we couldn't have them anymore.

One morning I woke up and I couldn't get out of bed because my back was hurting so bad. Every time I moved, it hurt. I told my family about my back, and my mom and sister tried to tell me to work somewhere else. But I decided to keep working at HP. I told my supervisor about my back, and she told me to assemble printers for two days. I still had to stand on my feet, but it gave my back a little rest.

After my second day of assembly, I was ordered to return to box load. My fellow workers and I started doing one hundred printers per hour, with three of us rotating, which kept us from breaking our backs. The HP management raised the target, and we had to meet it or the management would bitch at our Manpower supervisor. I was outraged she didn't speak out until I finally found out she was just a temporary worker like us. If she'd spoken out, chances are her temp assignment at the warehouse would have ended.

Since I was the youngest and fastest at box load, I was working extra hard. When one of my fellow workers had back problems, I'd tell them to do half and I'd make up the other half. I used to daydream about how HP was talked about at my high school—a place where you could really advance. The truth was that HP only did this for people with B.A. or B.S. degrees. The rest of us were to remain temps.

I left after eight months. I know now I won't do warehouse work or assembly work unless it's the last resort. My family has been in Silicon Valley for so long, I don't want to leave. I see my future here. I'm going to try to get my bachelor's degree and hopefully help make things change in the Valley.

EDWARD NIETO

BRAVE NEW FIELD

One afternoon, when our nation's security color code
went to orange, a commercial popped up on my TV screen
for homeland security training and jobs. It was from a
group called the National Institute of Technology (NIT),
and it really grabbed me.

In the NIT commercial, people in uniforms talked
about all the new jobs in homeland security. I thought that
maybe I could make money, build a career, and help out
our nation in a time of need—all without leaving loved
ones. I called and made an appointment with the recruiter
the next day.

The interview site in San Jose was packed mostly with
male applicants of color from around the Bay Area. The
recruiter explained that the school costs about $8,000 for
a seven-month program that prepares you to be a "Homeland Security Specialist." They had classes ranging from
"Tactical Communications" to "Domestic and International Terrorism" to "Emergency Planning and Security
Measures." The literature stated that the Homeland Security Specialist diploma program "helps prepare graduates for

careers in the security industry as corporate and government security and safety personnel." I asked the recruiter
if I could get a government job and maybe even become a
spy. She said, "Yeah, this would be a good place to start."

I found out that the school used to focus on computers,
but now it's all about security. The recruiter also emphasized that there were only two spots open.

I left with two packets—one describing the program's
courses and the other detailing how to get help with tui-
tion. On the cover of the financial aid packet was a picture
of this dude holding a wad of cash in his fist. The paperwork also included listings of the types of jobs popping up
in the homeland security industry. Jobs like "law enforcement," "border patrol," "Homeland Security Officers" and
"Critical Infrastructure Assurance Officer," not to mention
"Coast Guard civilian jobs" and "Customs Service."

The pay ranged from $12,776 to $142,498 a year, with
all jobs aimed at keeping America safe. The packet also
listed numerous private sector jobs. After looking more
carefully, though, I noticed that a lot of the jobs were just
downloaded from the Internet—stuff I could have gotten
on my own. Worse, the recruiter made it clear that the
school can't guarantee a job after the Homeland Security
Specialist degree.

After poring over the paperwork and spending a sleep-
less night, I came to the conclusion that this was not for
me—the cost alone was enough to deter me. When I went
back to the office and told the interviewer, she was dis-
appointed. She insisted that it would all be good with the
financial aid. "But I need cash, and I need a job now," I told
her, "not seven months from now."

Showing me a generic degree, she said, "Now doesn't this look nicer than your high school diploma?" Actually, I thought, if you changed the color and some of the words, it looked exactly like my high school diploma. And since there are people with Ph.D.s out there who are unemployed and desperate, getting any of these new homeland security jobs would be very competitive, if not impossible, for someone like me. I asked her what kind of homeland security jobs she could hook me up with now. She brought in an employment specialist who told me what jobs were out there, and how to go about finding them. She said that they would give me information for a security job at Target if I enrolled.

I declined. Driving home, I concluded that the only thing I would gain from the program would be the ability to say that I'd gone to school. As for an underappreciated security job for eight dollars an hour at Target, I think I can get that on my own.

ANGEL LUNA

TRADE SCHOOL SCAMS

You know those TV commercials about how to easily get the security of a diploma at a trade school? They really do work—the commercials at least.

When I kept seeing them, I was really interested in the advertised school. Not because I thought it was cool, but based on how quickly you could finish and get a job in that field. Like many of my friends, instead of going to a community college I chose to go to one of these vocational schools after high school.

I didn't do it because I thought I wasn't smart enough to go to a regular college. At the time that I enrolled, I saw trade school as the quickest and most efficient way for me to start my career in graphic design. Many people my age were enrolling in community college just to front that they were not some mediocre guy or girl with no dreams and a shady job. But after going to vocational school I still feel like that mediocre guy, except I also threw away $10,000 in doing so.

The representative or "counselor" that I met with told me wonderful stuff about the school. It was like she was describing Disneyland. The process was very easy and I was surprised that I got accepted according to their academic standards. My future looked bright according to them. I was a star in the making, soon to be on the wall of fame of Silicon Valley College.

Admission was a done deal already, so the next step was to get my financial aid. The financial aid lady was very nice, and while she stared at my paperwork, she kept telling me how wonderful it is to get an education.

She said that I should get my mom as a cosigner for my federal loan. I took the bus home with my application in my hand, thinking of the many possibilities in becoming a professional in a field that involves technology. But my mom could not cosign because of her legal status, and I also couldn't apply on my own, because I was not old enough. I was brokenhearted, but I was still trying to fight to get my education.

According to the school, there was a grant from the school for "special emergencies," and my emergency fit the requirements. I got the private grants, and in my eyes I was happy that there were still people out there trying to help a young man get his education.

The first day came around. The class was Photoshop for beginners, and I was very confident because I had already worked with the program. The teacher was very smart and relaxed. For the first time, I felt like a teacher cared. Plus, this guy was a pro in the industry—he actually made video games, something I always dreamed of. I really felt like I was in a good position to reach my goal of becoming a graphic designer.

Months went by, and I was there every day, on time and ready to do what I was told. It was great. The last day of

Photoshop came, and to my surprise all we had to do was hand in easy assignments. So the class just played a Street Fighter II tournament the rest of the day.

But after that first class, all the others had the same vibe: the students weren't being taught, and we were just having a good time. All the teachers were taking the same approach, just kicking back and trying to pass the time. Only we were being charged $1,000 per class.

I had to take time off for a short while to take care of some family issues, and when I got back, all the school personnel had changed. They weren't sure what to do with me, so I kept getting switched around the graphic department like a top.

Even while I was still taking classes there, I received calls and letters about paying back my loans to the school. Ultimately, that's all they really wanted from me anyway—my dollar bills. The only thing that I always received on time at that school was the reminder to pay my bills for the private loan, and let me tell you, those fools from Sally Mae are not that polite. I eventually just left the school after about a year.

The thing that broke my heart is that the school didn't care about the students. I experienced it. I would come late and no one would say anything, and I'd turn stuff in after deadlines and then they would give me an A or a B.

Because my loans are so big, I'm now working at UPS, and I have two other jobs. I've enrolled at DeAnza College. I didn't get the classes I wanted, but it's good to get used to the feeling of being in school again. After everything I went through, I'm not really stressing about getting into graphic design. I'm more concerned with getting a real education.

SHAMAKO NOBLE

THE ELEVATOR

I had been in Earthlink Business Sales for around a year.
It was one of the best jobs I'd ever had. For a variety of
different reasons. If you were at Earthlink at that time,
you were practically at a cultural center of San Jose. Every
Friday when we first arrived, there were keggers in the
downstairs lounge area.

Earthlink employees were offered a healthy balance
of both structure and freedom. At the beginning of each
month we were given our sales goals, and as long as we
reached them we had some flexibility in how we managed
our time.

So while everybody was about their money and business, everybody was about their fun and partying too. Most
of us smoked weed throughout the day, so there was a pretty good chance that if you were calling Earthlink, you were
getting someone who was keyed up. In addition, Earthlink
held space in the two tall buildings on Second and Santa
Clara in downtown San Jose. Right next to them was a bar
called Mission Ale House, which we affectionately referred
to as "Building Three."

One of the best parts of the job was the reliable and
trusty underground employee network. The very network

which was informing me that we were to be laid off soon. There were rumors the company was moving to some cheaper location. They'd already laid off customer service; over a hundred people, many of whom were homies of mine, laid off in one fell swoop less than six months or so ago. VPs arrived one random day asked us all to come to the lobby. My coworker referred to it as a mass grave.

At Earthlink, when VPs spontaneously showed up, and the whispers began, you knew that some dramatic change was going to take place. And it was more than likely that that change was going to result in more jobs or less jobs by some significant number. One time one of those visits had created the Broadband Installation Support Team. On another, it wiped out a whole floor of positions.

To be honest, I'd mentally checked out from the job a while ago. Granted, working at a call center is an honest living. But there was no soul in it. No proof that I was actually alive, breathing and being. It was mechanical, and I'd learned to engage with it as such.

Jerry was one of our top sales people. A white male in his midthirties, Jerry used to smoke weed with me in the parking lot, a tradition at Earthlink. We'd transferred over from Broadband Installation with each other. He'd come to dominate our sales department, and was recognized as one of the company's top salesmen. As a result, he was greatly rewarded with glass trophies, raises, and a sense of long-term security. He'd decided to get married, have a kid, and buy a house. From the treatment that he was receiving at Earthlink, that made a lot of sense.

On one particular day, we were told early on, as soon as our shift began, that there was a VP there, and that that VP

wanted to speak specifically to our department. We knew
that today could be our day to get the ax.

When we went to go see the VP, it was around 1:00 p.m.
We had one manager and two supervisors, both of whom
were named Scott. One Scott was a gentle giant.

He was around six-four and weighed in at a good 260
or above. He had a mustache and looked like a mountain
man, with the appropriate camping and fishing pictures in
his office. By contrast, the other Scott was a shorter, dark-
haired spark plug. He was always full of energy and could
get the team fired up in an instant. The color and size
contrast was always striking, but their spirit of both team
energy and good times was consistent with the vibe of the
whole office. They were all cool cats. By chance, I ended
up in the elevator with one of the Scotts, Jerry, and couple
of other folks. Jerry cut straight to the chase.

"Are we getting laid off?"

"No, not at all," Scott said. "Nothing like that."

I was pretty sure at that point that that's exactly what it
was. And I'm pretty sure that Jerry knew as well. In fact,
I'm pretty sure that everybody in the elevator knew, so
why he decided to keep it to himself when he was out-
numbered by awareness, I do not know. That question
came early in the ride, and while it was only four floors, it
was surely the longest elevator ride of my life.

The chopping block was sweet and simple. Nothing
compared to what some companies give, but a lot more
than others. Hell, at this stage of the game some companies
just up and disappeared. We received two months' salary
and all sick days and time off. If we wanted to, we were
offered the opportunity to transfer to Roseville or Pasade-
na, two entirely different lives than the one offered by San

Jose. I wasn't really bothered by all this, but Jerry was. He wasn't happy at all.

We had lunch, got our laid-off-or-relocate option, and headed back to our desks. Somehow, I ended up back in the elevator with Jerry and Scott 1.

"We're not getting laid off, huh?" Jerry said.

Scott 1 laid his head down in shame and spoke in his own defense the best way he could. " Look man.....I'm— I'm sorry. I didn't mean anything by it, I just—"

Jerry cut him off. "Yeah, I know. I get it. You're a company man. I know the deal."

That first elevator ride was nothing compared to this one. Even though, to be honest, I was happy to be leaving. I was also learning a valuable lesson on what happens when you base your life on these companies when you're just regular cats like us. Even the "Jerrys" of the world can come and go to them.

IRENE EDAN

I USED TO LAY PEOPLE OFF, THEN IT HAPPENED TO ME

Human resources is the cleanup crew of any company. It's the dumps, the dirty work. It's what your manager doesn't want to do. It was created with the impression of keeping the employees' best interest in mind, but what it's really for is to cover the company's ass from any legal liabilities. I started working at a large Internet marketing firm about a year and a half ago, looking for a more structured lifestyle. This was what I thought I needed. At the time, I wanted some kind of order…something that gave me a sense of security. A friend of mine worked at the same place I did and told me about a job opening they had. I ended up getting the job and never realized how getting it and losing it would impact my life.

I was somewhat thrown into the job. I started as a secretary, and due to my boss, the HR manager's neglecting of responsibilities, I soon became the HR assistant and mostly responsible for all HR issues in our building. I learned to do my job well, but also grew to hate it. As time passed I felt chained for someone of such a young age. At only nineteen, working and going to school, both full time, started to weigh hard on me. I felt part of the American natural order of things, when that has never been my way. I started to neglect old thought, old lifestyle, and pretty much everything I ever used to love. The thing about working in HR is that it consumes most of your life. Emotional investments are abandoned, stress begins to devour your mind and body, and you start to feel, in a sense, lost. Corporate welfare becomes your only concentration. The worst part was the constant layoffs.

Even though we worked mostly through agencies, and it was never my decision who stays and who goes, I still felt the weight of people's lives being thrown on my shoulders, so much that I literally started to get pains in my upper back, often feeling like I was carrying a small child on my shoulders. It was tough making those phone calls. Knowing that when an employee passed by my desk on their way out of the building saying, "Have a good day, Irene!" that it would be the last I would ever see of that person.

I remember a time that I had to give an exit interview to an employee. He was a good worker and, in fact, I was the one who got him the job there. This employee was my friend, and about a year before this I was grip-taping his skateboard at a local skate shop I worked at downtown. "Work is too slow right now, we cannot keep you. Please know that you were a great employee and we wish you our best regards. Unfortunately, your work here has ended."

Slow words fell from somewhere, which seemed was not my own mouth. Who was I?

I had a false sense of power that did not come from a high-position role, but was only given to me because no one else wanted it. In that sense, I was a bitch. I was the company's bitch, and because no one else wanted to own up to their decisions, it was my job to make sure they got done.

Emotionally, things got worse over time. Early last year when the economy really started to decline, I started reviewing overqualified resumes. I got people who had college degrees, years of experience in managerial positions, people who were pushed out of their previous jobs because of a declining economy and downsizing.

I'd look at them and think, "Why would you want to work on a production floor for a fifth of what you used to make?"

The answer was clear: they wanted to eat and they wanted their family to eat. Just like I want to eat, and just like I sucked it up and did what I needed to do to provide for my bills. I should have known that soon enough, even my role was not impervious to our nation's waning economy.

Companies always make you feel like you are part of them, like your job is completely secure. This is especially true in HR. So when I got laid off in January '09, this immediately came as a shock to me. I always knew who would be laid off at our company, and now I was the on the other end of the spectrum.

During my exit interview, their reasoning felt unjustified. I felt like I was being told, "We don't have to do this, but we are, just in case." It was humiliating. This is the same humiliation I felt when employees asked me why they had lost their jobs. Even though they had assured me that this was completely a business decision and it had nothing to do with my performance as an employee, I still felt like my job was wasn't important enough to keep. In the Internet economy, it's a dog-eat-dog business like any other, and I should have known that even HR is not invincible.

So here I am now, with no job, but I still continue to go to school full time. I have no family support, and at a young age, am a very independent person.

Even though I have an added burden on my chest, as far as how my bills will get paid, I also feel a huge load has been taken off of my shoulders. I have never been able to do anything I ever wanted to do or considered doing because my job tied me down. I needed that job. Mostly because I needed the car I drove to get to my job, and in order to keep the car I had, to keep the job I had, I needed to work to pay for it. I dole my life to that car, and it was a huge robotic cycle.

If that formation was interrupted, I always felt I would experience dire repercussions. Now I'm at that point, and to be honest, I feel fine. In fact, I feel like now is a better time than any to change to route I'm traveling. Now is the time to start a new chapter and take advantage of the blessings in disguise I'm given. I have always known that I did not want this job; I just never had the courage to do anything about it. Someone else did it for me.

TOM DEARBORNE

FINDING WORK AFTER THIRTY– FOUR YEARS IN PRISON

My last paycheck behind the walls was for nineteen cents an hour. I was a sought-after employee, having served thirty-four years and already proven my work ethic. Anytime a position came open in the prison, I had a significant chance of being assigned, if I chose to be. As my time came closer in February 2014 for Governor Brown to allow the parole board to release me, or not, I had the expectation that—if I were to be released—I would have to work a *little* harder than the average job seeker, but that I would be gainfully employed in a couple of weeks at the latest.

My expectations were completely unrealistic. Ahead of any serious effort at job hunting, I had to navigate the maze of DMV, social security, and the various other appointments my parole agent had made for me. Before being incarcerated in 1980, I could walk into a business and

FINDING WORK AFTER THIRTY-FOUR YEARS IN PRISON

talk my way into a job that might very well start the same
day. Now I was being told to apply online and, "Somebody will get back to you." It made me wonder if a person
would ever see it.

All of the Lifer support groups I attended in prison
emphasized being honest about a record so there would be
no negative repercussions later. What they failed to understand is that employers for the jobs I was applying for only
checked back seven years. I could put "no" in the box and
still be right, as my last arrest was 1980. It cost me my first
two opportunities for work. I tried to explain that as a Lifer, I had to earn my way out and make significant changes
most prisoners weren't required to. I thought the employer would feel my excitement and recognize I had proven
myself many years ago. That wasn't the case. As I shared my
journey and the joy at being able to work for real amounts
of money, I could see their eyes glaze over and their body
postures close up. Clearly they weren't viewing what I had
overcome with as much enthusiasm as I had hoped.

Job fairs, online ads, and networking seemed to be a slow
process. After all, I just wanted to work. Why couldn't they
see I was just the guy they needed? Finally, my opportunity
came and I was ready for it. I work a full-time station at
a new steakhouse and grill. After the position was mine, I
realized how anxious I had been about finding work, paying
my bills, and making the adjustments necessary to function
in this new world. It is a relief that allows me to sleep better
at night and welcome the new day with confidence.

My current struggle is one of communication. Almost
everyone in the kitchen speaks Spanish, and while I can get
by—it's just barely. Even my supervisor (who speaks much
better English than I speak Spanish) clearly uses English

as a distant second language. Some of my friends grumble about them not learning English, but I take a different view. If I want to do well there, I need to do my part, not expect everyone to change for me. Estoy aprendiendo Español muy despacio. I think of what it must be like when a Spanish speaker is trying to learn a job with nothing but English speakers, and I can now truly empathize.

Empathy? Me? Finding a desire and ability to empathize is one of the measuring tools I use to keep me on the forward path. I feel grateful for this chance to be a member of the community, and I'm willing to take whatever steps are necessary to do it well. Finding employment has been key for me to evolve from a surreal feeling and wondering if I belong here, to feeling confident and hopeful about the future. Employment is a huge factor in stabilizing a long-term parolee. It brings with it the feeling of having a life preserver in the ocean while treading water.

There are other pieces to the puzzle that various groups stepped up with as I entered this bright new world from prison on February 12, 2014. The Re-Entry Center, Community Solutions, Good Samaritan, and Bridges of Hope all helped my transition in ways I'll always remember and pay forward.

I feel grateful each and every day and take my place in society seriously. I want to be a good citizen, a good neighbor, and a good man. You'll find that most of the prisoners who have served more than two or three decades in the California prison system have similar sentiments. These people need employment soon after release. That investment will reap this community untold amounts of return by people who are some of the most motivated and appreciative people you will ever meet.

ALEX GUTIERREZ

REBIRTH THROUGH FINDING A NEW JOB

While sitting in the café lobby of a brand-new tech building in North San Jose, I signed my name on the last piece of paper of an employee packet for my new job as a dishwasher at a national catering company. It felt good to be able to shake the boss's hand after I filled out the forms, or even have a boss to shake hands with once again. I feel reborn.

Since March I have been led on a wild goose chase for the perfect full-time job. I have traveled across the state, even went outside of the state pursuing odd jobs of varying tasks—I've worked in event setup, was a security guard, and even a traveling soap salesman. After a year, I was still lost and confused, and ultimately frustrated with the economy, employers, and myself.

It wasn't easy getting this job. I had been staying on the streets, had to get cleaned up at a public restroom downtown, and had to travel three hours using my last five dollars, just hoping to land it. But I got it.

After leaving the building of my future livelihood, on the light rail back to San Jose from Milpitas, I couldn't help but to look back and take a glance in disbelief. I finally got another opportunity to have full-time employment, and even have medical and dental benefits, something very few people my age (at twenty-three years old) have through their work. Having a job means I can really live again. When I say live, I mean actually live a decent life of renting my own pad again, buy some new clothes, pay for a decent meal, and finally be able to enjoy things without having to be dependent or having to wait on someone to help me with a handout.

What the high unemployment statistics don't show is the direct hit mentally, even nervous breakdowns, that not having a job can have on a person. Unemployment can steal all of the faith you had, right from inside of you.

After a few days into the new job, I also made a new friend at work. We got along, and he was nice enough to even let me stay at his house with his family. While staying there, I noticed a jump in lifestyles. What I have thought of as privileges for the past few years is pretty much normal day-to-day life for others. I was so used to sleeping on the street that even in the apartment I slept with all my clothes and shoes on. It took a while to know that I can actually fully sleep, and not have to worry if anyone was trying to harm me or take any of my belongings, that I could really just rest. Unemployment becomes psychological because it forces you to create a completely different lifestyle than what you once had, and you can get used to that lifestyle.

I noticed what I had gotten used to as someone without income when I got my first check. We went shopping and I felt I had splurged a little bit on myself—buying some new clothes. It had been so long, I didn't know any of my sizes, and was scared to put them on because I didn't want to get them dirty. I was terrified of putting them on because I wasn't sure of how long it would be until I would receive some new clothes. It was a habit from the streets, so it took a while for it to sink in that I was working full time, making decent money, and I could buy some more clothes and afford to wash them.

I've been working at my new job for a few months now and am getting used to the stable living. The work may be tedious manual labor, but every two weeks I look at my paycheck and smile. After a hard day's work I get to go home to take a shower, watch TV, or just chill.

With this new job, I feel as though this is the start of a new foundation for my life with a grand opening. It's like watching a building being built. My foundation has been built on an empty lot with bad piping and bad soil, with sewer water running everywhere. But with every new layer of foundation laid down comes a purification from the new piping and new soil.

My first level of my construction has already started with my housing situation becoming stabilized, and with starting a bank account. Even though I have only twenty-five bones in my checking account, it sure feels good to have twenty-five—it's better than nothing at all.

I still have friends that are lost in the mix of unemployment, doubting their own capabilities of getting back into the workforce, school, or society for that matter. The feeling of rejection can haunt anyone, and scare them away from an opportunity that could be theirs if they reach out for it. Those feelings really just interfere with the greatness that we all know deep down inside we can accomplish.

HOUNG NGUYEN

THE WORLD OF MA'S SALON

My ma owns this hair shop downtown called La Petite Salon. I've been working here for over a year now, and have watched my ma work here for ten years. It's a small, sugar-coated hole-in-the-wall, and most of the clientele are Vietnamese women. Since I was born and raised in the United States, it's a place where I can get closer to the Vietnamese culture and community.

This shop's four walls endure, preserve, and hold in all the emotions of every woman who has ever walked in. That's why a lot of these women come, in fact—to leave with something more than a cut or color. They're seeking self-assurance, comfort, and laughs.

All my ma's clients are referred to her by word of mouth. It's funny, because I swear I'll meet a lady one day and next thing I know I've met all the females in her family. Lately it's been real slow. Not to the point where we have to close shop, but compared to before, where she'd be booked all week, we're slow. I'm not experienced enough to feel the anxieties that fall on her, but I see ma always waiting for a call on her cell from a client. But she still keeps the salon going with all her skills. I'm proud of her, you know what I mean? She's just crazy, though.

Instead of a client coming in every month or two for a trim or cut, now we won't see them till three, four, or five months later. Some ladies would rather walk around with three to five inches of untouched roots than pay to cover or fix it. It's funny, because everyone is saying they just want to grow their hair long, but you know the real reason for the style change.

Just the other day, I was blow-drying this woman's hair, Chi Truc, who's in her late twenties. She started talking to me and my ma about how she's been unemployed for about five months now. She said she's just pretending it's a vacation she deserves after all of her years of being a workaholic. I could tell by her voice she was scared. Then I saw how she felt more at ease after my ma told her the success stories of other clients she's had lately.

We've got regulars from all occupations—doctors, lawyers, real estate agents, factory workers, cooks, and housewives. We got the old lady, Ba, who always wants a perm and an elegant do when she walks out, like she's got someplace fancy to go to. We got the party girl, Co Nga. Now this lady is fly, and probably in her late thirties. She dresses in trendy name brands and parties hardy. She makes nasty jokes, has attitude when time calls for it, and is always gossiping. Co Hien never talks, just nods, and says "Thank you" when my ma's done. She seems plain and simple, but sad.

It's hella funny when the subject of men, hubby or boyfriend, comes up. It's like the signal to all the women in the shop that it's okay to speak their minds. From under the dryers the heads pop up with, "Girl, I feel you," and "Honey, I been there, done that, and about to recycle some more." My ma listens, makes jokes, and gives advice where she feels fit. She's heard just about every kind of story, so the ladies respect her feedback. I just listen and laugh. They always say, "Just wait, honey, you got a long way to go," or, "Always think for your family and yourself first."

The shop makes me feel connected to these Vietnamese women and the culture. I feel rooted and happy to work alongside my ma and hear all these life stories, knowledge, and teachings. It's hard work standing and doing things all day in the shop, but I love what I do, who I do it with, and who I do it for.

LIZ GONZALEZ

THE CONNECTED SPIRIT

Lessons from a Silicon Valley meditation guide.

The heart is the guide
No other law to abide
A journey that is only yours to take
Life creates with those awake
When the focus is only on the outer
Distractions, they grow louder
The stress makes a mess of your head
Struggling without an end
The true richness of each one is denied
Exalting material, although it's belied
To fit in amongst clones
Yet, we are more than skin on bones
Shallow, hollow, no—no longer
Now's the time to open wide
Throw all the illusions aside

Reveal the true nature inside
A glowing heart of pink and gold
Legends from the time of old
A spirit connected to one and all
Answer to your highest call
Your heart is the guide
No other law to abide

ABOUT THE EDITORS

RAJ JAYADEV is the coordinator and cofounder of Silicon Valley De-Bug. He resides in San Jose, California, with his wife, Charisse, and son, Nanjappa Bakani Jayadev.

JEAN MELESAINE is a Samoan American documentary photographer based out of Oakland, California. Working with Silicon Valley De-Bug for over ten years, she was taught conscious photography from community members as a teenager. Her parents are from the villages of Moamoa and Faleali'li in Western Samoa.

About Heyday

Heyday is an independent, nonprofit publisher and unique cultural institution. We promote widespread awareness and celebration of California's many cultures, landscapes, and boundary-breaking ideas. Through our well-crafted books, public events, and innovative outreach programs we are building a vibrant community of readers, writers, and thinkers.

Thank You

It takes the collective effort of many to create a thriving literary culture. We are thankful to all the thoughtful people we have the privilege to engage with. Cheers to our writers, artists, editors, storytellers, designers, printers, bookstores, critics, cultural organizations, readers, and book lovers everywhere!

We are especially grateful for the generous funding we've received for our publications and programs during the past year from foundations and hundreds of individual donors. Major supporters include:

Advocates for Indigenous California Language Survival; Anonymous (3); Arkay Foundation; Judith and Phillip Auth; Judy Avery; Carol Baird and Alan Harper; Paul Bancroft III; The Bancroft Library; Richard and Rickie Ann Baum; BayTree Fund; S. D. Bechtel, Jr. Foundation; Jean and Fred Berensmeier; Joan Berman; Barbara Boucke; Beatrice Bowles, in memory of Susan S. Lake; John Briscoe; David Brower Center; Helen Cagampang; California Historical Society; California Rice Commission; California State

Parks Foundation; California Wildlife Foundation/California Oak Foundation; Joanne Campbell; The Campbell Foundation; Candelaria Fund; James and Margaret Chapin; Graham Chisholm; The Christensen Fund; Jon Christensen; Cynthia Clarke; Community Futures Collective; Lawrence Crooks; Lauren and Alan Dachs; Nik Dehejia; Topher Delaney; Chris Desser and Kirk Marckwald; Lokelani Devone; Frances Dinkelspiel and Gary Wayne; Doune Trust; The Durfee Foundation; Megan Fletcher and J.K. Dineen; Michael Eaton and Charity Kenyon; Richard and Gretchen Evans; Friends of the Roseville Library; Furthur Foundation; The Wallace Alexander Gerbode Foundation; Patrick Golden; Erica and Barry Goode; Wanda Lee Graves and Stephen Duscha; The Walter and Elise Haas Fund; Coke and James Hallowell; Theresa Harlan and Ken Tiger; Cindy Heitzman; Carla Hills; Leanne Hinton and Gary Scott; Sandra and Charles Hobson; Nettie Hoge; Claudia Jurmain; Kalliopeia Foundation; Judith Lowry and Brad Croul; Marty and Pamela Krasney; Robert and Karen Kustel; Guy Lampard and Suzanne Badenhoop; Thomas Lockard and Alix Marduel; Thomas J. Long Foundation; Bryce Lundberg; Sam and Alfreda Maloof Foundation for Arts & Crafts; Michael McCone; Nion McEvoy and Leslie Berriman; Moore Family Foundation; Michael J. Moratto, in memory of Major J. Moratto; Stewart R. Mott Foundation; Karen and Thomas Mulvaney; Richard Nagler; National Wildlife Federation; Native Arts and Cultures Foundation; The Nature Conservancy; Nightingale Family Foundation; Steven Nightingale and Lucy Blake; Northern California Water Association; Panta Rhea Foundation; Pease Family Fund; Jean Pokorny; Jeannene Przyblyski; Steven Rasmussen and Felicia Woytak; Susan Raynes; Restore Hetch

Getting Involved

To learn more about our publications, events and other ways you can participate, please visit www.heydaybooks. com.